BEETLE CONSERVATION

BEETLE CONSERVATION

Editor

T. R. New

La Trobe University, Melbourne, Australia

Reprinted from *Journal of Insect Conservation*
Volume 11, Number 1 (2007)

Springer

A C.I.P. Catalogue record for this book is available from the library of Congress.

ISBN-13 978-1-4020-5987-2 (HB)
ISBN-13 978-1-4020-6047-2 (e-book)

Published by Springer
P.O. Box 17, 3300 AA Dordrecht, The Netherlands

www.springer.com

Cover illustration: Coral Pink Sand Dunes Tiger Beetle (Cicindela albissima). Photo by Christine Breton

Printed on acid-free paper

Table of Contents

J Insect Conserv (2007) 11:1–4
DOI 10.1007/s10841-006-9022-0

EDITORIAL

Beetles and conservation

T. R. New

Published online: 1 November 2006

Many biologists claim that beetles are the most diverse of all animal groups, with the number of described forms commonly cited as around 350,000. Whether or not they will prove eventually to be outnumbered in species by Hymenoptera (with their vast array of tiny parasitic taxa still largely not enumerated), Coleoptera are undoubtedly enormously rich in species and widespread in many terrestrial and freshwater environments throughout the world. As a consequence of the widespread knowledge that Charles Darwin was a beetle enthusiast as a young man, of Haldane's comment about the likings of the Creator, and of Erwin's (1982) focus on tropical beetles to estimate wider species abundance, as well as around two centuries of hobbyist and collector interests, beetles are amongst the most popular insects. Their richness is acknowledged widely as valuable both per se and in wider evaluations of ecological condition. Beetles are important also in a variety of other applied and more esoteric contexts. Ball-rolling scarabs were venerated in ancient Egypt as an analogue of the sun-god (Ra) rolling the sun across the heavens. Beetles have long been attractive to collectors, with rare species (or large 'trophy individuals' of taxa such as stag beetles [Lucanidae]) sometimes commanding enormous sums—but one outcome of this interest is a critical mass of interest and accumulation of sound substantial taxonomic and ecological information on a wide variety of taxa, particularly in the northern hemisphere, together with information on responses of beetles to environmental changes. Thus, studies on responses of beetles to habitat fragmentation in many parts of the world have given important ecological insights of much wider relevance in practical conservation: see papers in this journal by Lövei and Cartellieri (2000, New Zealand Carabidae) and Trumbo and Bloch (2000, North American Silphidae), as examples.

T. R. New (✉)
Department of Zoology, La Trobe University, Victoria 3086, Australia
e-mail: T.New@latrobe.edu.au

Some beetles are decidedly 'charismatic', and a number of the more distinctive such groups have been promoted widely in conservation. The tiger beetles (Carabidae: Cicindelinae—sometimes treated as a separate family, Cicindelidae), for example, have become sufficiently popular to merit a specialist journal 'Cicindela', founded in 1969) and, following impetus suggested by Pearson (1988), have become a significant global flagship group for beetle conservation, with numerous recent studies on the management of individual species, particularly in Europe and North America. Many beetles command attention as major pests of crops or stored products, with some, conversely, valuable as predators to suppress a wide range of pest arthropods. Efforts to promote the wellbeing of carabids and others as native predators through conservation biological control have been important contributions to understanding the management of habitats for beetles, and have much wider conservation implications. The designs of 'beetle banks' and conservation headlands, for example, have much wider relevance as refuges and reservoir habitats in highly altered landscapes (New 2005, for references).

Evans and Bellamy (1996) used the term 'beetlephilia' (extending from Wilson's [1994] 'biophilia') to summarize the fascination of these insects to many people and, in one of very few general essays to do so, emphasized the values and needs to conserve beetles as important components of the biosphere, with numerous keystone species and sustainers of numerous terrestrial ecosystem services. Freshwater beetles also can be important 'indicators', and members of about 11 families are obligately marine, most of them confined to intertidal or littoral zones (Doyen 1976). More generally, beetles are divided amongst about 160 families, with around two-thirds of all species allocated to only about 8 of these. The largest family, Curculionidae s.l. (the weevils), contains around 50,000 described species and the other large families (Buprestidae, Carabidae, Cerambycidae, Chrysomelidae, Scarabaeidae, Staphylinidae, Tenebrionidae) are all widespread terrestrial groups.

Beetles are important in conservation. Their long representation in the fossil record (facilitated by hardness and persistence of their elytra and other diagnostic structures in more modern deposits) has given them a unique role in helping to interpret long-term faunal changes, particularly for Quaternary assemblages. Their immense ecological variety leads to meaningful information on changes in richness and composition of local assemblages with changing conditions, and studies on pest species and their natural enemies have strengthened the framework of knowledge available for conservation studies. Beetles are among the most frequently cited species of insects as conservation targets. A number of species have undoubtedly become extinct as a result of human activities, but we have little idea of how many. Sixteen taxa are listed as extinct in the 2004 IUCN Red List of threatened species. The largest component of these (6) is from Hawaii, followed by the continental United States (3) and New Zealand (2), but this number seems almost certainly to be a severe underestimate. We have no real idea of the extent of continuing centinelan extinctions over much of the tropics, for example. Mawdsley and Stork (1995) recorded only 10 reported extinctions of Coleoptera, all of them from islands, and with the New Zealand taxa probably exterminated there by predation from introduced rats. These authors, in a wider examination of reductions in British beetles, found a substantial variety of threats, with various threats differentially affecting different beetle groups and their habitats. The early, 1988, IUCN Red List of threatened animal species listed at least 350 beetle species (uncertainty because of listings such as 'all species of genus X'), even more than for butterfly species at that time. More recent changes have lessened this number somewhat, but additions include a series of *Colophon* (Lucanidae) from South Africa which are apparently threatened by over-collecting: 'beetlephilia' can be a two-edged sword in conservation! But the passions beetles arouse, together with the interests and concern they foster, may assuredly be an asset in promoting conservation awareness.

Concerns for beetles, and exploration of their values in wider environmental appraisal have involved studies ranging from single species (with some becoming notable flagship taxa in various parts of the world) to assemblages. Such studies continue to diversify, and to contribute to a wider and more informed spectrum of activities in insect conservation. This special issue brings together some recent information on beetle conservation studies, ranging from overviews to specific cases. They are a small fraction of current activities, but indicate a variety of ways in which beetles are contributing to wider conservation, as foci of specific conservation programmes and as organisms with values in broader environmental assessments in many parts of the world. And, as with other insects in conservation, it is pertinent to consider the varying levels of capability and interest in different places. For the United Kingdom, with (as for butterflies and others) one of the best documented and understood beetle faunas on Earth, the fine filter species level approach to beetle conservation is well-advanced, with a number of species (such as the stag beetle, *Lucanus cervus*) important flagships for conservation endeavour. Species Action Plans exist (at March 2006) for 87 individual species and for three "groups" (river shingle beetles, *Harpalus* spp., saproxylic beetles). The last of these draws on longer-term concerns for persistence of dead wood and the need for continuity of mature timber habitats, and includes also the need for construction of artificial habitats for rare saproxylic Elateridae, and includes themes in common for 10 species. Concerns for saproxylic beetles have resulted in a series of European Symposia on the theme (see Barclay and Telnov 2005), with concerns evident also in North America and Australia. The Species Action Plans for British beetles are short documents, including brief statements on current status, factors causing loss or decline, current actions, action plan objectives and targets and proposed actions with lead agencies. They thereby specify the major trends and threats and note what is needed, and who is to have responsibility for undertaking the necessary conservation actions. In contrast, the recovery plans for several of the dozen beetles listed under the United States Endangered Species Act are much more detailed: thus the plan for the American burying beetle (*Nicrophorus americanus*, the first-listed beetle [1989] under the Act) occupies 81 pages (USFWS 1991), and that for the Puritan tiger beetle (*Cicindela puritana*), 47 pages. As for Britain, some North American beetles have become flagships for particular threatened habitats. The Delta Green ground beetle (*Elaphrus viridis*) is one of a suite of species (including 10 plant species of concern) highlighting the significance of Vernal Pool Ecosystems in California, for example. Captive breeding programmes for notable beetles have been instigated increasingly in Europe and North America, but are still relatively novel elsewhere.

Beetle conservation in much of the tropics inevitably lacks such detailed focus, although particular species of collectable scarabaeoids, in particular, have been signaled as probably threatened by exploitation. Several such species are protected formally in Taiwan, for example. However, few species-orientated programmes have so far been implemented. The Seychelles Giant Tenebrionid beetle (*Polposipus herculeanus*) is confined to Frégate Island, and is one of few to become the subject of a captive breeding endeavour and field monitoring programme. South Africa, Australia and New Zealand represent an intermediate situation, where some concerns over the better-documented groups of beetles have arisen, but the perspective is largely incomplete. Terrestrial examples from South Africa and New Zealand are included in this special issue, and the following brief comment on knowledge and conservation profile of Australian beetles complements these.

Documentation of Australian Coleoptera is highly incomplete at the species level. As with most larger orders of insects in Australia, there are few resident specialists on Coleoptera, but appreciation of the values of beetles for appraising both freshwater and terrestrial communities has led to interest from biologists with wider priorities in seeking to establish the effects of land management practices on our invertebrate fauna. However, many of the enumerated species (of an estimated 28,300 beetle species: Lawrence and Britton 1994) are undescribed, and numerous families have not been revised recently so that evaluations such as the listings for some larger groups in various volumes of the Zoological Catalogue of Australia series (namely Adephaga, Lawrence et al. 1987; Scarabaeoidea, Cassis et al. 1992; Elateroidea, Calder 1998; Buprestoidea, Bellamy 2002) inevitably contain numerous specific epithets of unknown validity, rendering the identification of many species particularly difficult for non-specialists without easy access to a major institutional collection. A progressive series of illustrated keys to genera of South Australian beetles (Matthews 1982–1997; Matthews and Reid 2002) is a useful adjunct in the south east, and more local synoptic treatments, such as to the jewel beetles of Tasmania (Cowie 2001, in which a number of taxon names are flagged as 'tentative', pending wider generic revisions), also augment more general summaries such as that by Hawkeswood (1987). Nevertheless, for many groups of beetles there is little practical option but to operate at the 'morphospecies level' in analyzing assemblages or working toward effective rapid biodiversity assessments. In general, Australia's beetles have received relatively little specific conservation attention as conservation targets, other than concerns for some 'collectable taxa' such as Buprestidae and Lucanidae (with European trade data for these and Scarabaeidae summarized by Hawkeswood et al. 1991, who noted the predominance of Buprestidae [with at least 59 species, many of them endemic to Western Australia, in this trade during the previous decade]), and the impacts of forestry practices on the latter, particularly in Tasmania, see Meggs and Munks (2003, Lucanidae) and Michaels and McQuillan (1995) and Michaels and Mendel (1998, both on Carabidae). Altogether, only about 50 species, in 12 families, had been signalled (some of them with little objective information) as of individual conservation concern by 1997 (Yen and Butcher 1997). Earlier, Hill and Michaelis (1988) listed four beetles amongst their 'select list' of threatened insects, with the major threats signaled as 'urbanization' (three species: a scarabaeid (*Aptenocanthon rossi*) and two species of *Stigmodera*, Buprestidae, all from New South Wales) and 'woodchipping' (one Tasmanian chrysomelid, *Chrysophtharta philomela*). Carabidae have not attained the high conservation profile they possess in other parts of the world. In their more recent summary, Clarke and Spier-Ashcroft (2003) noted only nine species listed individually for formal protection under either state or federal legislation, but this category includes also the controversial 'all species of Buprestidae' for Western Australia (below). This listing is likely to be a severe underestimate of the actual number of threatened species of beetles, but reflects the incomplete nature of the documentation by which invertebrates are still treated in Australia. Even fewer taxa have been studied in any detail, and the nature and levels of threat can largely only be inferred. Thus, the blanket listing of all jewel beetles (Buprestidae) as protected fauna in Western Australia (under the Wildlife Conservation Act) in August 1978 appeared to be motivated by suspicions of overcollecting for commercial use. However this listing did not prevent the almost simultaneous clearing of large patches of prime habitat for agricultural development in the state. Combination of habitat loss and commercial attractiveness for particular beetles, as with the tropical rainforest stag beetle *Phalacrognathus muelleri* in Queensland, may be a potent collective adverse influence for rarer species of such ‘collectable’ taxa. There is little doubt that the conservation status of numerous Australian beetles, both in terrestrial and freshwater habitats, is far more parlous than the limited information available implies. There has not, for example, been any serious attempt to evaluate the conservation needs of most larger radiations of Australian beetles, so that objective information on such enormous groups as Curculionoidea (some 6,000 species) and the phytophagous Chrysomelidae (about 3,000 species) is almost wholly lacking, other than for sporadic references to ‘rarity’. Similar comments apply to beetle faunas in many other places.

The papers in this special issue exemplify a variety of contexts and concerns in beetle conservation. They emphasize perspectives gained outside western Europe, where much of the most detailed work on beetle conservation has been undertaken. I am grateful to the authors for their enthusiastic participation, and to the reviewers of all papers for their prompt and constructive comments; comments by Michael Samways on a draft of this introduction have been welcomed. The support of Zuzana Bernhart and Pauline Lichtveld at Springer is also appreciated greatly.

References

Barclay MVL, Telnov D (eds) (2005) Proc. 3rd Symposium and Workshop on the Conservation of Saproxylic Beetles, Riga/Latvia 2004. Latvijas entomologs, Suplementum VI

Bellamy CL (2002) Coleoptera: Buprestoidea. Zoological Catalogue of Australia. vol 29:5. CSIRO, Melbourne/Australian Biological Resources Study, Canberra

Calder AA (1998) Coleoptera: Elateroidea. Zoological Catalogue of Australia. vol 29:6. CSIRO, Melbourne/Australian Biological Resources Study, Canberra

Cassis G, Houston WWK, Weir TA and Moore BP (1992) Scarabaeoidea. Zoological Catalogue of Australia, vol 9. Australian Government Printing Service, Canberra

Clarke GM and Spier-Ashcroft F (2003) A review of the conservation status of selected Australian non-marine invertebrates. Environment Australia, Canberra

Cowie D (2001) Jewel beetles of Tasmania. A field naturalist's guide. Tasmanian Field Naturalists' Club Inc., Hobart

Doyen JT (1976) Marine beetles (Coleoptera excluding Staphylinidae). In: Cheng L (ed), Marine insects. North Holland/American Elsevier, Amsterdam and New York, pp 497–519

Erwin TL (1982) Tropical forests: their richness in Coleoptera and other arthropod species. Coleopterists Bull 36:74–75

Evans AV, Bellamy CL (1996) An inordinate fondness for beetles. Henry Holt, New York

Hawkeswood T (1987) Beetles of Australia. Angus and Robertson, North Ryde

Hawkeswood T, Callister DJ, Antram F (1991) Collection and export of Australian insects. TRAFFIC Bull 12:441–448

Hill L, Michaelis FB (1988) Conservation of insects and related wildlife. Occas. Pap. no 13. Australian National Parks and Wildlife Service, Canberra

Lawrence JF, Britton EB (1994) Australian beetles. Melbourne University Press, Melbourne

Lawrence JF, Moore BP, Pyke JE, Weir TA (1987) Coleoptera. Archostemata, Myxophaga and Adephaga. Zoological Catalogue of Australia, vol. 4. Australian Governement Printing Service, Canberra

Lövei GL, Cartellieri M (2000) Ground beetles (Coleoptera: Carabidae) in forest fragments of the Manawatu, New Zealand: collapsed assemblages? J Insect Conserv 4:239–244

Matthews EG (1982–1997) A guide to the genera of beetles of South Australia. Parts 1–7. South Australian Museum, Adelaide

Matthews EG, Reid CAM (2002) A guide to the genera of beetles of South Australia. Part 8, Chrysomelidae. South Australian Museum, Adelaide

Mawdsley NA, Stork NE (1995) Species extinctions in insects: ecological and biogeographical considerations. In: Harrington R, Stork NE (eds) Insects in a changing environment. Academic Press, London, pp 321–369

Meggs JM, Munks SA (2003) Distribution, habitat characteristics and conservation requirements of a forest-dependent threatened invertebrate Lissotes latidens (Coleoptera: Lucanidae). J Insect Conserv 7:137–152

Michaels KF, McQuillan PB (1995) Impact of commercial forest management on geophilous carabid beetles (Coleoptera, Carabidae) in tall wet Eucalyptus obliqua forest in southern Tasmania. Aust J Ecol 20:316–323

Michaels K, Mendel L (1998) Carabid beetles and vegetation associations in the Tasmanian Eastern Tiers: implications for conservation. Pacif Conserv Biol 4:240–249

New TR (2005) Invertebrate conservation and agricultural ecosystems. Cambridge University Press, Cambridge

Pearson DL (1988) Biology of tiger beetles. Ann Rev Entomol 33:123–147

Trumbo ST, Bloch PT (2000) Habitat fragmentation and burying beetle abundance and success. J Insect Conserv 4:245–252

USFWS (United States Fish and Wildlife Service) (1991) American burying beetle (Nicrophorus americanus) Recovery Plan, Washington, DC

Wilson EO (1994) Biophilia. Harvard University Press, Cambridge, Mass

Yen AL and Butcher RJ (1997) An overview of the conservation of non-marine invertebrates in Australia. Environment Australia, Canberra

J Insect Conserv (2007) 11:5–18
DOI 10.1007/s10841-006-9014-0

BEETLE CONSERVATION

The effects of forestry on carabid beetles (Coleoptera: Carabidae) in boreal forests

Jari Niemelä · Matti Koivula · D. Johan Kotze

Received: 18 November 2005 / Accepted: 20 March 2006 / Published online: 1 November 2006

Abstract As compared to natural forests, managed boreal forests are younger, more homogeneous in terms of tree age and species composition, and consist of smaller fragments. Here we examine the effects of such characteristics caused by forestry on carabid beetles (Coleoptera, Carabidae) in the boreal region. The main results are the following. (1) Fragmentation of forests and the size of a fragment appear not to be crucial for the survival of the majority of forest carabids, as they tend to be distributed over various successional stages, but species requiring old-growth habitats suffer. (2) For carabids there appear to be no or very few edge specialist species, and forest-open land edges appear to be effective barriers for species associated with forest or open habitat. However, generalist species easily cross the edge, and edges of forest fragments may be invaded by species from the surrounding open habitat. (3) Habitat change following clear-cutting dramatically changes the composition of carabid assemblages: species restricted to mature forests disappear and open-habitat species invade, while habitat generalists survive at least in the short term. Carabid diversity can probably best be maintained if forest management mimics natural processes, maintains natural structures and includes the natural composition of vegetation and other structural elements (such as dead wood) within the stands, provided that these forest features can be maintained and recreated through forest management practices. At a larger scale, the whole spectrum of forest types and ages (especially old-growth forests), and different successional processes (especially fire) should be maintained. These require the development and use of innovative logging methods, and the planning, implementation, and assessment of landscape-scale ecological management strategies.

Keywords Carabidae · Forest management · Boreal forests

J. Niemelä (✉) · D. J. Kotze
Department of Biological and Environmental Sciences, University of Helsinki, P.O. Box 65, Viikinkaari 1, FI-00014 University of Helsinki, Finland
e-mail: jari.niemela@helsinki.fi

M. Koivula
Department of Renewable Resources, University of Alberta, 4-42 ESB, Edmonton, AB T6G 2E3, Canada

Introduction

Forestry is simplifying tree species composition, homogenising tree age structure, and fragmenting forested landscapes in many parts of the vast boreal region (Bonan and Shugart 1989). As a consequence, managed forests are younger overall, consist of smaller fragments with more edge habitat, and are exposed to novel disturbance regimes. For example, the proportion of old-growth forest in state-owned land decreased from 44% to 7% between 1910 and 1990 in Sweden (Linder and Östlund 1992), and forestry rather than wildfire is the main stand-replacing factor initiating secondary succession (Esseen et al. 1997). Similar changes have occurred in many North American boreal forests (Loope 1991).

These changes have had clear impacts on the boreal biota. Managed forests host fewer forest-specialist species and lower abundances of many common species than do unmanaged old-growth forests (here, forests in which the dominant trees are older than 150 years and with no management for at least the past 100 years). Examples include, in addition to carabid beetles (Coleoptera, Carabidae) discussed here, bryophytes and epiphytic lichens (Andersson and Hytteborn 1991; Kuusinen 1994, 1996; Dettki and Esseen 1998), beetles living in and on dead wood (Väisänen et al. 1993; Siitonen and Martikainen 1994; Jonsell et al. 1998) and mycetophilid flies (Økland 1994).

Carabid beetles are among the best studied taxa regarding the effects of forest management on the boreal biota. Despite their general abundance it has often been difficult to model the habitat requirements of these beetles. Thus, in several studies cited in this paper carabids have been divided into "forest", "generalist" and "open habitat" species. However, as is evident from Table 1 and the baseline literature (e.g. Lindroth 1985, 1986, 1992), such a classification is context specific. For instance, the term "forest" can be understood as meaning any area where trees grow, thus including e.g. clear-cuts, recent burns and other regenerating stands. Thus, a comparison between agricultural land and various treed areas (from clear-cuts to mature forest) may suggest that *Calathus micropterus* (Duftschmid) is a forest species, even a specialist, as in the data presented in Table 1 only 5% of the specimens were caught from agricultural habitat. However, in many studies "forest" is defined in a narrower sense: a treed area with a closed canopy and dominant trees being at least a few decades old. The younger regenerating phases, on the other hand, represent "open habitat". From this perspective, *C. micropterus* is classified differently. Table 1 shows that this species is equally abundantly caught from clear-cuts (logged 1–3 years ago) and unlogged mature forests, suggesting a forest/open-habitat generalist life style. To conclude, species classifications according to their habitat associations should be done by carefully considering the nature of the compared habitats.

As the focus of this review is on forests, we use the term "forest" to indicate a treed area with closed canopy and (nearly) mature trees, whereas younger successional phases are referred to as "open" habitats. Accordingly, Table 1 shows a literature and field-data based classification of most of the abundant southern Finnish forest-associated carabid species to "forest", "open habitat" and "generalist" species.

In this review we evaluate the effects of forestry on individual carabid species and communities in the boreal region. We will also identify threats, and propose ways to ameliorate the adverse effects of forestry on forest-dwelling carabid beetles.

Effects of forestry on carabids

Forestry practices often result in the fragmentation of the original forest landscape. Habitat fragmentation is the partitioning of continuous habitat into small remnants (e.g., Saunders et al. 1991; Hunter 1996; Haila 1999; Debinski and Holt 2000). Habitat loss is often included into the concept of fragmentation, and it is usually difficult to separate the effects of habitat loss from that of fragmentation per se on the biota. Here, we first focus on the effects of fragmentation, i.e. isolation, size, the configuration of the habitat remnants, and edge effects. Thereafter, we discuss the consequences of habitat loss on carabid beetles, and examine how species with different habitat affinities are affected.

Fragmentation

Forest fragmentation affects carabid assemblages in a myriad of ways (Niemelä 2001). For example, Niemelä et al. (1988) compared different-sized fragments of coniferous forest surrounded by clear-cut areas in Finland and found that although there were no differences in species richness, the abundance of many generalist and open habitat species was higher in small (< 5 ha) than in large (>30 ha) forest remnants. In other studies, species richness increased with decreasing size of forest patches surrounded by agricultural land (Niemelä and Halme 1992; Halme and Niemelä 1993): small fragments (0.5–3.0 ha) had 18 species, larger ones (9.6–21.5 ha) had 13 species, while continuous, unfragmented forest had only 11 species. These differences are most likely attributed to the small fragments being more open and grassier, and thus more favourable for generalist and open habitat carabid species primarily residing in the surrounding agricultural grasslands or open clear-cut sites. Similar results have been reported from countries in other biogeographical regions, e.g. England (Usher et al. 1993), France (Fournier and Loreau 2001) and Hungary (Magura et al. 2001). In addition, Usher et al. (1993) noted that the number of species was affected by the shape of the forest patch so that remnants with high edge-to-area ratios contained more species, probably because of higher invasion rates from the matrix.

Contrary to the above studies, asphalt-isolated Finnish forest patches of 0.2–1.8 ha in size hosted fewer species than larger ones (0.5–37.4 ha) (Koivula and Vermeulen 2005). Similarly, a study in the Canadian temperate region showed that carabid species richness increased with increasing forest fragment size (Burke and Goulet 1998). The smallest fragment with a forest core area of 0.3 ha had only two carabid species,

Table 1 Literature and field-data based division of some Finnish boreal forest-associated carabids into habitat affinity groups

Species	Habitat, literature		Habitat, field data (only for species with $n > 20$)			Habitat association	
	Forest	Open	Farmland vs. forest (% in forest)	Clear–cut vs. mature forest (% in mature)	Young vs. old regener. (% in older)	Literature assoc.	Field-data assoc.
Agonum fuliginosum (Panzer, 1809)	X	X	(98) 37.7	(326) 31.0	(114) 78.1	G	G
Agonum sexpunctatum (Linnaeus, 1758)	–	X	–	(173) 1.7	–	O	O
Amara brunnea (Gyllenhal, 1810)	X	–	(27) 100.0	(51) 74.5	–	F	F (G)
Amara communis (Panzer, 1797)	X	X	(55) 58.8	–	–	(G) O	G (O)
Amara eurynota (Panzer, 1797)	–	X	(107) 4.3	–	–	O	O
Amara lunicollis (Schiödte, 1837)	–	X	(22) 8.1	(162) 0.6	(67) 1.5	O	O
Anisodactylus binotatus (Fabricius, 1787)	–	X	–	–	–	O	–
Badister bullatus (Schrank, 1798)	X	X	–	–	–	G	–
Bembidion bruxellense Wesmaël, 1835	–	X	–	–	–	O	–
Bembidion guttula (Fabricius, 1792)	–	X	–	–	–	O	–
Bembidion lampros (Herbst, 1784)	–	X	–	–	–	O	–
Bembidion quadrimaculatum (Linnaeus, 1761)	–	X	–	–	–	O	–
Bradycellus caucasicus (Chaudoir, 1846)	–	X	–	–	–	O	–
Calathus erratus (Sahlberg, 1827)	–	X	–	–	–	O	–
Calathus melanocephalus (Linnaeus, 1758)	–	X	–	–	–	O	–
Calathus micropterus (Duftschmid, 1812)	X	–	(811) 95.1	(3313) 51.5	(634) 93.2	F	F (G)
Carabus cancellatus (Illiger, 1798)	–	X	–	(11) 9.1	–	O	O
Carabus glabratus (Paykull, 1790)	X	–	(38) 88.2	(218) 68.8	(72) 47.2	F	G (F)
Carabus granulatus (Linnaeus, 1758)	X	X	(140) 6.3	–	–	G	O
Carabus hortensis (Linnaeus, 1758)	X	–	(878) 91.3	(129) 95.3	(21) 90.5	F	F
Carabus nemoralis (Müller, 1764)	X	X	(122) 46.7	–	–	G	G
Carabus nitens (Linnaeus, 1758)	–	X	–	–	–	O	–
Carabus violaceus (Linnaeus, 1758)	X	–	–	–	–	F	–
Cicindela campestris (Linnaeus, 1758)	–	X	–	(55) 3.6	–	O	O
Clivina fossor (Linnaeus, 1758)	–	X	–	–	–	O	–
Cychrus caraboides (Linnaeus, 1758)	X	–	(43) 82.4	(98) 72.4	–	F	F (G)
Dyschirius globosus (Herbst, 1784)	–	X	–	–	–	O	–
Harpalus affinis (Schrank, 1781)	–	X	–	–	–	O	–
Harpalus laevipes Zetterstedt, 1828	X	X	(58) 86.4	(79) 19.0	–	G	G
Harpalus latus (Linnaeus, 1758)	–	X	–	–	–	O	–
Harpalus rufibarbis (Fabricius, 1792)			(119) 8.3	–	–	O	O
Harpalus rufipes (Degeer, 1774)	–	X	(409) 2.5	–	–	O	O
Harpalus tardus (Panzer, 1797)	–	X	–	–	–	O	–
Leistus ferrugineus (Linnaeus, 1758)	X	X	–	–	–	G	–
Leistus terminatus (Hellwig, 1793)	X	–	–	(20) 55.0	–	F	G
Loricera pilicornis (Fabricius, 1775)	X	X	–	–	–	G	–
Notiophilus aquaticus (Linnaeus, 1758)	–	X	–	–	–	O	–
Notiophilus biguttatus (Fabricius, 1779)	X	–	–	(126) 57.9	–	F	G
Notiophilus germinyi Fabricius, 1863	–	X	–	–	–	O	–
Notiophilus palustris (Duftschmid, 1812)	X	X	–	–	(20) 5.0	G	O
Patrobus assimilis (Chaudoir, 1844)	X	X	(122) 5.0	(49) 30.6	(35) 94.3	G	G (F)
Patrobus atrorufus (Ström, 1768)	X	X	(961) 29.6	–	–	G	G (O)
Platynus mannerheimii (Dejean, 1828)	X		–	–	–	F	–
Platynus assimilis (Paykull, 1790)	X	–	–	(20) 50.0	–	F	G
Poecilus cupreus (Linnaeus, 1758)	–	X	(402) 7.4	(29) 0.0	–	O	O
Poecilus versicolor (Sturm, 1824)	–	X	(69) 4.4	–	(27) 0.0	O	O
Pterostichus adstrictus Eschscholtz, 1823	–	X	–	(73) 0.0	–	O	O
Pterostichus crenatus (Duftschmid, 1812)	–	X	–	–	–	O	–
Pterostichus diligens (Sturm, 1824)	X	X	–	(24) 4.2	(65) 9.2	G	O
Pterostichus melanarius (Illiger, 1798)	X	X	(3453) 24.2	–	–	G	G (O)
Pterostichus niger (Schaller, 1783)	X	X	(1883) 39.7	(198) 36.4	(399) 6.3	G	G (O)
Pterostichus nigrita (Paykull, 1790)[a]	X	X	–	(22) 0.0	–	G	O
Pterostichus oblongopunctatus (Fabricius, 1787)	X	–	(352) 82.7	(2189) 34.9	(239) 35.6	F	G
Pterostichus strenuus (Panzer, 1797)	X	X	(26) 39.8	(34) 14.7	(30) 16.7	G	G (O)
Synuchus vivalis (Illiger, 1798)	X	X	–	–	–	G	–
Trechus rivularis (Gyllenhal, 1810)	X	–	–	–	–	F	–

Table 1 continued

Species	Habitat, literature		Habitat, field data (only for species with $n > 20$)			Habitat association	
	Forest	Open	Farmland vs. forest (% in forest)	Clear–cut vs. mature forest (% in mature)	Young vs. old regener. (% in older)	Literature assoc.	Field-data assoc.
Trechus rubens (Fabricius, 1792)	X	X	–	–	–	G	–
Trechus secalis (Paykull, 1790)	X	X	(466) 36.0	(374) 25.4	(118) 54.2	G	G
Trichocellus placidus (Gyllenhal, 1827)	–	X	–	(51) 0.0	–	O	O

"Forest" means treed areas with closed canopy except in farmland-forest comparison; "open habitat" refers to regenerating but not yet closed stands (unless otherwise specified). The first two columns show whether a given species is most often found in forests or in open phases of succession (a generalist is indicated with an "X" in both columns) based on Lindroth (1985, 1986, 1992) and Koivula (2001). The next three columns show results from three field data sets: (number of individuals in parentheses), % indicates proportion in closed canopy forests. The columns are (1) farmland vs. forested habitat, including clear-cuts (a total of 96 samples; percentages calculated from adjusted catch; data from Koivula, Niemelä & Kotze unpubl.); (2) mature spruce forest vs. recent clear-cuts (8 replicates for both habitat types; data from Koivula 2002a); and (3) young (5–10 years since clear-cutting) vs. old (30–60 years since clear-cutting) clear-cuts (8 replicates for both age classes; data from Koivula et al. 2002). The two last columns show habitat classifications based on the literature and on field data: F = species is associated with closed canopy forests; G = species is abundantly found in both open and closed phases of succession; O = species is associated with open phases of succession; letters in parentheses indicate alternative classifications for a few ambiguous cases

[a]Data include a possibility for *Pterostichus rhaeticus* Heer

while there were 16–17 species in fragments >30 ha, and 22 species in continuous forest. However, isolation confounded the results of Burke and Goulet (1998), as the lowest species richness was found in small, isolated fragments, while the species richness of less isolated small fragments was comparable to that of large fragments. Yet, some studies report no clear effects of fragmentation on carabids. Abildsnes and Tømmerås (2000) concluded that experimentally fragmented forests (40 m × 40 m and 150 m × 150 m clear-cut sites) did not result in clear area effects and that isolation affected forest-dwelling species only slightly. Time since fragmentation may be a possible reason for this apparent lack of response: sampling was done only one year after clearing the forest (Abildsnes and Tømmerås 2000).

Simply counting the number of species, however, provides little information on the specific effects of fragmentation. Species richness used as a measure of conservation value may be misleading because disturbances may favour widespread and abundant generalists, leading to increased species richness as can be the case for carabids (e.g. Niemelä 1997). In contrast, analysing the responses of individual species is more likely to provide an understanding of the processes that lead to the extinction or persistence in fragmented landscapes (Davies and Margules 1998; Ewers and Didham 2006). For example, Koivula and Vermeulen (2005) showed that the response of *Pterostichus oblongopunctatus* (F.) to forest patch size in southern Finland depended on habitat quality that was indicated by tree-species composition and field-layer vegetation.

Based on current evidence, forest fragmentation and the size of a fragment appear not to be crucial for the survival of the majority of forest carabids, as they tend to be distributed over various successional stages (i.e. they are generalists; Ås 1993; Davies and Margules 1998; Gandhi et al. 2001). However, sensitive species requiring certain structural elements of old-growth forests, such as abundant large-sized dead wood, lush patches of mesic vegetation or wet spruce mires, or micro-climatic conditions created by closed canopy, suffer from forest fragmentation. In a Finnish study, many forest species were absent from small (0.5–3.0 ha) fragments, probably because of a lack of suitable habitat (Halme and Niemelä 1993). Similarly, Bauer (1989) and Usher et al. (1993) reported that individual specialist species responded positively to fragment size so that their species richness increased with increasing fragment size. Likewise, in Halme and Niemelä's (1993) study the strictest of forest species, most notably *Carabus glabratus* Paykull and *C. violaceus* L., occurred only in continuous forests. These species are flightless, which probably hampers their movements across grassland surrounding the studied forest patches (see Riecken and Raths 1996).

Edge effects

Fragmentation leads to a decreasing proportion of interior and an increasing proportion of edge habitat (e.g., Murcia 1995; Haila 1999; Matlack and Litvaitis 1999). Changes in abiotic and biotic conditions at the edge are called 'edge effects'. These include changes in wind, humidity, radiation, predation, parasitism and species interactions (Saunders et al. 1991; Andrén 1995; Chen et al. 1995; Murcia 1995; Donovan et al. 1997; Laurance 2000). The extent of these effects is strongly influenced by edge structure. For example, Didham and Lawton (1999) showed that edge effect penetration distances of many variables of microclimate and vegetation structure can be as much as five times greater at open edges lacking sheltering vegetation than at closed, densely vegetated edges.

It has been shown that forest edges harbour a rich invertebrate fauna (e.g. Helle and Muona 1985; Jokimäki et al. 1998) that is a mixture of species found on either side of the edge zone (Kotze and Samways 1999). However, species requiring interior habitat suffer from edge effects (Stevens and Husband 1998) and, consequently, such species may be lost if fragments become too small and dominated by edge habitat (Haila 1999). To find ways to counteract the negative effects associated with edges it is important to understand how species respond to conditions at habitat edges (Haila et al. 1994).

For boreal carabids there appear to be no or very few edge specialist species (Spence et al. 1996; Abildnes and Tømmerås 2000; Heliölä et al. 2001; Koivula et al. 2004). Similar observations have been made in tropical systems (Didham 1997) and in the Mediterranean region (Taboada et al. 2004). Very small forest patches (a few tens of m^2, or 20–40 m wide forest strips) within clear-cuts have been shown to host a similar carabid fauna to clear-cuts (Koivula 2002a), i.e. such patches do not represent forest for carabids.

Some reports indicate that edges of forest fragments are easily invaded by species from the surrounding matrix, and some species may continue through the edge 'filter' into the forest interior (Spence et al. 1996; Koivula et al. 2004). Contrary to this, Heliölä et al. (2001) found that carabid assemblages in mature spruce forests vs. clear-cuts were fairly clearly separated by the edge, with open-habitat species being restricted to the clear-cut.

As regards the distribution of forest species across edges, Koivula et al. (2004) showed that forest species rarely entered the field, and if they did, only to a maximum distance of 20 m. For farmland species the forest edge was even more abrupt, as very few individuals crossed the edge, and they tended to avoid the edge zone by decreasing in abundance 5–10 m from the forest patch.

For the conservation of intact forest habitat, a crucial question is how far into the forest interior the edge effects reach. Open-habitat carabids are reluctant to enter the forest (Heliölä et al. 2001), and if they do, they usually penetrate only tens of metres into the forest (Spence et al. 1996; Koivula et al. 2004). Thus, to the extent that it is possible to generalise from these studies, it appears that carabid communities of forest fragments of reasonable size (a few ha) remain more or less intact, and that the edge functions as an efficient barrier that stops open-habitat species from invading the fragment interior (see also Hänggi and Baur 1998; Kotze and Samways 1999). Edge effects thus apparently do not make boreal forest fragments vulnerable to invasion by native open-habitat carabids.

Although native open-habitat carabids do not seem to penetrate forests, alien invasive species may be an exception, and therefore of concern for the conservation of native carabid assemblages. For example, Niemelä and Spence (1999) reported that the introduced *Pterostichus melanarius* Ill., an open-habitat to generalist species in Fennoscandia (Lindroth 1986), was spreading from road verges into mature forests in Canada.

Habitat loss and change

Carabid species richness and the abundances of generalist and open-habitat species usually increase following forest cutting (Niemelä et al. 1988, 1993a, b; Halme and Niemelä 1993; Haila et al. 1994; Spence et al. 1996; Beaudry et al. 1997; Heliölä et al. 2001; Koivula 2002a, b; Sippola et al. 2002; Pearce et al. 2003; de Warnaffe and Lebrun 2004). The open, dry and warm clear-cuts are favoured by many carabid species usually found in grasslands and other similar habitats (e.g. genera *Harpalus* and *Amara*), while only a few species are associated with the dark and cool spruce forest (Niemelä 1993; see also Table 1). In addition to carabids, clear-cuts are also rich in species of spiders (Pajunen et al. 1995), ants (Punttila et al. 1991, 1994), butterflies (Väisänen 1995) and sun-loving dead-wood-dependent beetles (Martikainen 2001).

Clear-cutting affects the abundance of forest-associated species negatively (Niemelä et al. 1993a, b; Langor et al. 1994; Spence et al. 1996; Werner and Raffa 2000). However, such species may occur in recently clear-cut sites for a few years (Koivula 2002a, b). This observation has at least three explanations. First, individuals of forest species caught in the clear-cut may be wanderers from nearby mature stands (Spence et al. 1996). This may be the case, for example, in sites close to mature-stand edges (Koivula 2002b; Koivula et al. 2002). Second, catches of forest species may represent populations on their way to local extinction (or drastic decrease). Because some carabids may live for at least 2–3 years (van Dijk 1996), these species can persist for some time in the clear-cuts. For example, populations of *Carabus hortensis* L., *Cychrus caraboides* L., *Platynus mannerheimii* Dej. and *Calathus micropterus* (Dft.) survived at least 2–3 years in Finnish clear-cuts (Koivula 2002a, b) but either decreased or disappeared later on (Niemelä et al. 1993a; Koivula et al. 2002; see also Abildsnes and Tømmerås 2000 and Table 1). Third, some species with a more generalist life style may be able to tolerate natural disturbances (e.g. windthrows and forest fires) better and may, therefore, survive in clear-cut sites.

In clear-cut sites, the 'pioneer' carabid assemblages gradually change as the succession of vegetation proceeds and the relict populations of forest species disappear. A drastic change usually occurs at canopy closure that takes place in the boreal region ca. 20–30 years after clear-cutting (Niemelä et al. 1996; Koivula et al. 2002; Koivula and Niemelä 2002; see also Table 1). After canopy closure, open-habitat species decline rapidly and species requiring mature forest become dominant.

Interestingly, the closed-canopy boreal forest appears to be a rather hostile environment for many carabid species, as only a few species are able to thrive (Koivula et al. 2002), resulting in peculiar species-abundance distributions in which there is a gap between the dominant species and the scarce ones (Niemelä 1993). For carabids, therefore, boreal forests may be an adversity or A-selected environment (Greenslade 1983) with low species richness (but not necessarily low densities). Such specialist species are rather few in the boreal forest making up < 10% of the total forest carabid fauna (Niemelä 1997). Today, these species are in danger because of forestry.

Forest specialists are sensitive to habitat changes caused by forestry

Most forest-dwelling carabids require specific habitat elements (Niemelä et al. 1996; Koivula 2001; Koivula and Niemelä 2002; de Warnaffe and Lebrun 2004; Work et al. 2004), with many species showing associations with certain micro-site characteristics (Heliölä et al. 2001; Koivula 2002a, b; Koivula et al. 2002). These species can be considered forest-habitat specialists. For example, the circumboreal spruce-mire specialist *Platynus mannerheimii* Dej. requires wet sites dominated by *Sphagnum* mosses within mature coniferous forest (Niemelä et al. 1987).

Another important factor affecting carabid distribution is the amount of leaf litter, indicating the importance of scattered deciduous trees within coniferous-dominated stands (Niemelä et al. 1992; Koivula et al. 1999; Heliölä et al. 2001; Koivula et al. 2002; see also Poole et al. 2003; Magura et al. 2005). Deciduous-tree admixtures, together with spruce mires, are also important for epiphytic lichens, land snails and dead-wood-dependent beetles (Kuusinen 1996; Esseen et al. 1997; Niemelä 1997; Martikainen 2000; Siitonen and Saaristo 2000).

Due to their specialised habitat requirements, species associated with old-growth forests are at risk in the modern managed forest landscape. In the Palaearctic region, such carabid species are few but include, for example, *Carabus hortensis* L., *Cychrus caraboides* L. and *Platynus mannerheimii* that are typically found in mature coniferous forest (Koivula 2001, 2002a; see also Niemelä et al. 1993a and Table 1). To demonstrate the responses of such species to management, Koivula and Niemelä (2002) studied the responses of several species to micro-site variation, forest edges and logging, and presented a graphic representation depicting examples of the responses of a few forest carabid species to forestry (reproduced in Fig. 1; for a classification of these species, see Table 1). This model is rather similar to that of Spence et al. (1996) but with two major differences: firstly, open-habitat species do not colonise large forest stands (Heliölä et al. 2001), and secondly, the majority of forest species are assumed to survive in the clear-cut sites adjacent to mature forest stands.

According to this model some species, such as the forest species *Calathus micropterus*, suffer from clear-cutting and decrease in abundance (Fig. 1). Also the abundance of *Carabus hortensis* is lower in the clear-cut than in mature forest (Koivula 2002a), but individuals best persist in specific micro-sites, such as retained tree groups or in mesic and luxuriant clear-cut sites (not shown). These two species thus vary remarkably in abundance along the successional gradient but persist over the early open phases in low abundances (Koivula and Niemelä 2002). The latter species may demand lush vegetated patches to survive, however. The model also includes an effect of distance from mature forest. *Calathus micropterus* decreases in abundance toward clear-cut central areas (Fig. 1). Open-habitat species, such as *Pterostichus adstrictus* Eschz., on the other hand, show an opposite trend by increasing in abundance in the clear-cut with distance from mature forest edges (Heliölä et al. 2001; Koivula and Niemelä 2002; Fig. 1). Some species are forest generalists with varying degrees of micro-habitat requirements. For example, *Agonum fuliginosum* (Pz.) appears to be a generalist with respect to canopy closure, but is dependent on *Sphagnum* mires or other moist micro-sites (Fig. 1; see also discussion on *C. hortensis* above). Thus, if any mire habitat remains

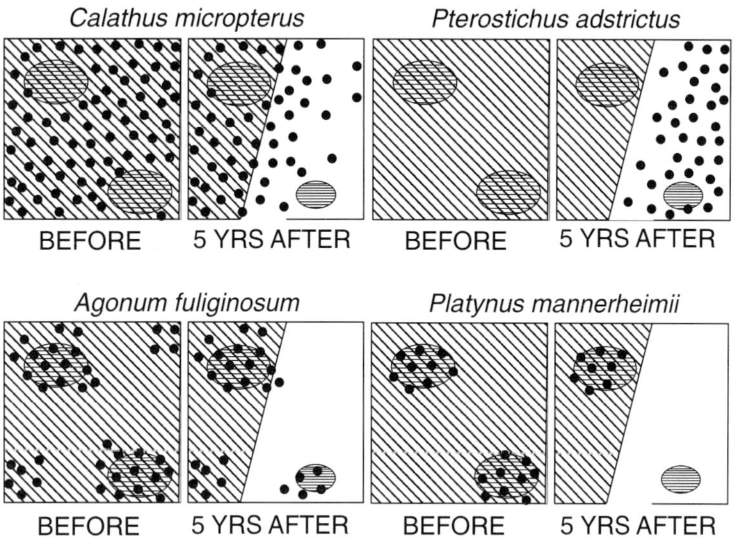

Fig. 1 Graphic model for four boreal forest carabid species showing their responses to micro-site characteristics (here, spruce mires), edge distance and clear-cutting within 1-ha forest blocks. The situations before and 5 years after partially clear-cutting are shown. Shaded sections are unlogged mature spruce forest (*diagonal lines*) or spruce mires (*horizontal lines*), white areas are clear-cuts, and black dots indicate the presence of a given species (more densely placed dots indicate higher abundance). For species classification and model interpretation, see the text

in the clear-cut, *A. fuliginosum* will also persist there. Another mire specialist, *Platynus mannerheimii*, on the other hand, disappears from the clear-cut, as it seems to require closed canopy forest (Niemelä et al. 1987, 1993a, Fig. 1).

Modified harvesting methods have a lower impact on carabids

In recent years, harvesting practices have changed in the boreal forests of Fennoscandia and Canada (e.g. Hallman et al. 1996; Angelstam 1997; Schneider 2002). Various new practices intended to be environmentally sound have been developed (e.g. Vanha-Majamaa and Jalonen 2001), all with the intention of leaving more live and dead wood in the forest. The question is—do these practices better maintain the biota of mature forests than does clear-cutting?

One of the alternative logging methods in Fennoscandia is single-tree thinning and gap felling, mimicking wind-throws. In the short term, thinning and the creation of such small gaps (3 openings/ha, each 30–50 m in diameter) only moderately changed the forest carabid assemblage, while traditional clear-cutting and retention felling (2–3 groups of 10–30 retained trees/ha) caused more profound changes (Koivula 2002a, b; Koivula and Niemelä 2002, 2003; see also Atlegrim et al. 1997). The proportion of retained trees (70%–90%) in these thinned and gap-felled stands may provide enough canopy cover even for the more sensitive forest species. From a longer-term study in the temperate forests of Oregon, Peck and Niwa (2004) reported that no differences in total abundance of carabids or species richness were found between stands thinned 16–41 years prior and unthinned stands, indicating that thinning has only minor effects on carabid communities. Overall, gap felling has been shown to maintain the original assemblages better than large clear-cuts (Koivula 2002a).

Logging methods form a continuum from clear-cutting and retaining small groups of trees (large changes in carabid assemblages) to thinning and gap felling (smaller changes in carabid assemblages). This pattern is highlighted by examining the response of open-habitat carabids to canopy cover (Fig. 2). Here, control (unlogged) and thinned stands (with ca. 400–1000 >1.3-m tall trees/ha) only rarely hosted open-habitat carabids, whereas gap felling (ca. 50% of trees removed in three 40 m × 40 m patches per ha) hosted considerably more individuals. Retention felling (with 10–30 trees in 3 groups/ha) and clear-cutting host these carabids abundantly (Fig. 2), indicating stronger changes compared to logging methods with more retention.

Moreover, Vance and Nol (2003) reported that carabid communities in stands harvested by single-tree-selection cutting 15–20 years previously contained a species composition similar to un-logged mature stands, suggesting few long-term effects on the carabid biota after the first rotation. Furthermore, Moore et al. (2004) showed that selective cutting and strip clear-cutting had no significant impacts on species diversity

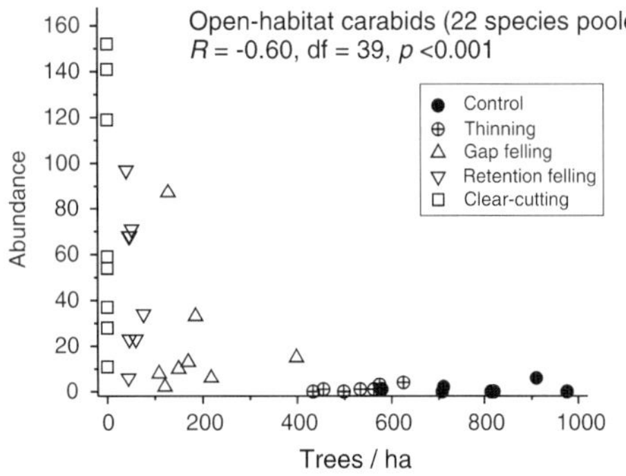

Fig. 2 The abundance of open-habitat carabids, plotted against a gradient of an increasing amount of retained trees (*horizontal axis*). Sampling was done in 40 differently logged spruce-forest stands in south-central Finland during second and third post-harvest summers. Logging methods were control (unlogged), thinning, gap and retention felling, and clear-cutting. Samples from the same stand over the 2 years were pooled. Data from Koivula (2002a, b)

and richness 6–13 years after treatment, yet some species were more abundant in either the managed forests or the control sites. Thus, retention of components that help sensitive species to persist in the managed landscape and avoiding homogenisation of stands are preferred management practices from the point of view of biodiversity (Law and Dickman 1998; Lindenmeyer and Franklin 2002). Preliminary estimates from Finland indicate that the harvesting costs per harvested timber unit in gap felling and retention felling are not much higher as compared to traditional clear-cutting.

Above we have demonstrated that forestry practices indeed affect carabid assemblages, but that the intensity of these practices determines the rate of change. A critical question remains: how persistent are these changes following forestry practices (Niemelä 1999)? More specifically, do carabid assemblages recover after timber harvesting? If they do, and all species re-colonise the harvested sites some time during succession, we may be able to maintain carabid assemblages of natural forest by long enough rotations and by leaving source areas for colonisation. If, however, recovery does not take place, forest tracts close by and of reasonable size (at least a few ha) should be left intact to provide habitat for forest-dwelling carabids. What evidence is there to illuminate this question?

A Canadian study supports the recovery view by showing that carabid assemblages in stands selectively logged 15–20 years prior to the study had started to recover towards assemblages in mature stands (Vance and Nol 2003). Also, Heyborne et al. (2003) reported indications of recovery, as beetle assemblages (including carabids) paralleled changes in plant communities from young herb-stage regrowth to mature forest during a 17-year study. Evidence is, however, equivocal as Niemelä et al. (1993a, b) reported that carabid assemblages did not recover even after 25–30 years post-harvest in western Canada. However, Koivula et al. (2002) showed that the carabid assemblages changed drastically 20–30 years since clear-cutting in Finland. Stands 60 years of age hosted virtually no open-habitat carabids, and the carabid assemblage closely resembled faunas reported from mature forests. Thus, it remains unclear whether Niemelä et al. (1993a) studied the carabid assemblages over a long enough chronosequence to document recovery. Furthermore, differences between the studies may be due to the history of forestry activities in these regions. Heavy forest use over a long period of time (e.g. in Finland) may have resulted in the most sensitive forest species having disappeared (see Kotze and O'Hara 2003), causing less changes and a rapid recovery in carabid assemblages after current harvesting as compared to sites where pristine forest in being logged, e.g. western Canada (Niemelä et al. 1993a).

The role of spatial scale for forest carabids

The maintenance and recovery of mature-forest species in the managed forest landscape requires attention at two interlinked spatial scales, viz. landscape and stand scale (Work et al. 2004). At the landscape level, conservation of a range of forest types and age classes is required to maintain the diversity of ground-dwelling beetles (Werner and Raffa 2000). For strict old-growth species, such as *Platynus mannerheimii*, reasonable amounts of old-growth stands should be retained to act as reservoirs and sources of colonists (Spence et al. 1996; Niemelä 1997, 1999; Werner and Raffa 2000). Also, several other species associated with mature forests benefit from retaining intact closed-canopy forests, e.g. *Carabus hortensis* and *Cychrus caraboides*. Thus, such habitats must be present in the landscape.

The spacing of suitable habitat patches in the landscape is crucial for species with limited dispersal powers (Desender et al. 1999), such as *Carabus hortensis* and *Cychrus caraboides* (Lindroth 1985, 1986). Within a fragmented landscape, the dispersal abilities of species and the distance between favoured habitat patches are critical for their survival (den Boer 1990a, b; Fahrig and Merriam 1994; With et al. 1997; Hanski 1999; Niemelä 2001). Flightless carabids may move up to a few hundred metres by foot (Mascanzoni and Wallin 1986), but these movements are seldom target-oriented at the scale of tens of metres (Wallin 1986). Koivula (2001) and Koivula et al. (2002) reported that the abundance of forest-dwelling carabids decreased in clear-cuts and sapling stands as the size of the clear-cuts and its distance from adjacent over 40-year-old stands increased. These findings indicate that the spatial configuration of mature and maturing stands in the managed forest landscape is of great importance to forest-dwelling carabids.

The retained old forest stands should include specific structural features (Werner and Raffa 2000). These include coarse woody debris (Work et al. 2004) that is an important habitat for many carabids (Pearce et al. 2003; Koivula et al. 2005). Moreover, *Platynus mannerheimii* requires *Sphagnum*-covered patches within mature spruce forest to survive (Niemelä et al. 1987; Liebherr and Song 2002). At the stand level, the new logging methods that leave more trees in the harvested stands may favour some forest species because retained trees shelter the ground layer from direct sunlight and, to some extent, from microclimatic alterations (Jalonen and Vanha-Majamaa 2001). However, the sheltering efficiency depends on the number of trees retained, as indicated by the negative relationship between the abundance of generalist carabids and increasing tree density (Koivula 2002b), and the degree of exposure and geographical location of the site. The most sensitive sites in spruce forests, such as spruce mires, should be left intact or managed carefully. Forest management should also aim at minimising the adverse effects of edges on forest specialist species, for example by leaving large enough fragments for forest-interior specialists to persist (Spence et al. 1996; Burke and Goulet 1998).

Those open-habitat carabid species that thrive in forest openings, on the other hand, are not at risk in the boreal forests of today, as the large areas of clear-cut habitat and young sapling stands function as source habitats for them (e.g. Spence et al. 1996; Koivula 2006). In Finland, such species are commonly found even in isolated clear-cuts (over 200 m from the nearest clear-cut; Koivula 2002b). In addition to dispersal by flight, the dense forest-road network in Finland (Västilä and Herrala-Ylinen 1999; Martikainen 2000) enhances the colonisation of recent clear-cuts by open-habitat species by offering dispersal corridors (Koivula 2003, 2006).

Changing forestry to become compatible with the maintenance of carabid diversity

In this review we showed that there are, broadly speaking, three types of carabid beetle responses to forestry practices. In strongly managed stands, such as clear-cuts, (1) open-habitat species appear and increase in abundance (but disappear when the canopy closes ca. 20–30 years later); (2) forest generalists persist throughout the clear-cut originated succession; and (3) species requiring mature closed-canopy forests are affected negatively by management and may not recover within several decades. The latter group further seems to consist of two recovery types: (a) the majority of forest-specialist species tend to recover following logging. This view is supported by the low forest carabid abundances during the first ca. 20–30 years of clear-cut originated succession followed by an increase later on (30–60 years; Koivula et al. 2002), but (b) several species show poor or no re-colonisation even after tens of years after the harvesting event (Niemelä et al. 1993a).

As there are more open-habitat species that colonise clear-cuts than there are forest species disappearing from them (Niemelä et al. 1988, 1993a, b; Koivula 2002a, b), species richness tends to increase following clear-cutting or other major forestry practices. Modern forestry creates clear-cuts suitable for open-habitat and disturbance-tolerant species, and consequently these are the 'winners' in intensively managed forest landscapes today. Also forest-habitat generalists (species found in different types and ages of forest) appear to be thriving. From a conservation point of view, species requiring mature closed-canopy forest are of concern. Their habitat continues to become fragmented and lost.

What can be done to maintain populations and assemblages of mature forest carabids and other forest taxa in the managed forest landscape? Obvious approaches include the decrease of clear-cut size, adoption of logging methods other than clear-cutting, and retention of certain "key" biotopes. Swedish and Finnish management guidelines and forestry laws have recently changed towards an ecologically more sustainable direction (Angelstam and Pettersson 1997; Savolainen 1997; Karvonen 2000). In Finland, for example, clear-cut size is usually less than 3–4 ha (Parviainen and Seppänen 1994), above which the economical benefit/cost relationship does not increase markedly (Imponen and Kaila 1988). Management guidelines also include the retention of ecological corridors, the maintenance of certain key habitats, and the use of modified logging methods (Karvonen 2000).

Natural forest dynamics is increasingly used as a guideline when practicing forestry (Angelstam 1998). However, this approach relies on the assumption that natural variation in stand structure and composition can be adequately maintained and recreated through forest management practices (Work et al. 2004). This matching is a challenge as there are several critical differences between natural disturbances and timber harvesting (Niemelä 1999). For example, in the boreal region, it has been suggested that clear-cutting mimics wildfire as a stand-replacing disturbance. According to this logic, species adapted to wildfire would also be adapted to clear-cutting. However, this is only partly true in carabids because assemblages in recently burned stands include fire-dependent species that are lacking from or are scarce in harvested unburned stands (Holliday 1984, 1991, 1992; Wikars 1992, 1995; Beaudry et al. 1997; Niwa and Peck 2002; Koivula et al. 2005; Saint-Germain et al. 2005). Overall, McCullough et al. (1998) concluded that studies examining the effects of forest fire on boreal carabid beetles suggest that the effects of fire on species diversity are small, but species composition may change (see Wikars and Schimmel 2001). However, these effects may depend on fire characteristics: Koivula et al. (2005) showed that the abundance of the fire specialist carabid *Sericoda quadripunctata* (DeGeer) within burns of a post-fire mixed-wood forest in Canada is determined by fire severity (measured using tree survival and scorch in the studied stands).

Although all forestry practices apparently result in changes in carabid assemblages, maintaining patches of closed canopy of a minimum of a few ares per hectare retains mature-forest carabid assemblages better than traditional clear-cuts (Koivula 2002a, b; Koivula and Niemelä 2002, 2003). However, this practice increases the amount of edge habitat in the stand (and at larger spatial scales; see below). Also, uncut strips between small clear-cut openings represent edge habitat for species requiring forest interior habitat. Therefore, in addition to invading the small openings, open-habitat carabids also invade the uncut strips between the openings (Koivula 2002a; Koivula and Niemelä 2003). Retained trees and vegetation in logged sites may nevertheless act as a buffer for adjacent forest reserves by decreasing edge effects and increasing the effective area of interior reserves (Lindenmeyer and Franklin 2002).

Modifications of the gap felling method are worthy of further study. For example, how does the spatial arrangement of retention patches or cuts affect the carabid fauna? Moreover, strip clear-cutting might maintain forest species assemblages better than traditional clear-cutting, but we are not aware of carabid studies that compare these logging methods. Nevertheless, Moore et al. (2004) showed that 60-m wide uncut strips host different assemblages than adjacent 60-m wide harvested strips. Increasing the logging rotation considerably from the current 80–120 years may have positive consequences on the forest fauna. Some important old-growth structures, such as decaying wood, large trees and trees with cavities, may need 200 years or longer to develop after harvesting (Siitonen 2001). Furthermore, a central European study indicated that populations of forest-specialist carabids may need centuries to return after strong forest-habitat alteration activities (Desender et al. 1999).

At the larger spatial scale, i.e. up to hundreds of hectares, forest biodiversity can probably best be maintained if forest management mimics natural processes of regeneration, such as fire-initiated succession, and maintains the natural age and tree-species composition (Fries et al. 1997). These views are supported by studies on carabids. Forest management should clearly aim at maintaining natural landscape structure and processes, including the availability of old-growth forests (see above). The key here is to maintain large-scale stand variation together with careful management at the smaller spatial scales (within stands). Moreover, Beaudry et al. (1997) suggested that the presence of regenerating sites with and without prescribed burning, in addition to preserved mature forest, could contribute to the maintenance of carabid diversity and to the avoidance of the extermination of some sensitive old-growth species. A challenge for forest managers is the introduction of fire to the North European boreal forests. This is imperative based on the current rarity of several fire specialists, such as the carabid *Sericoda bogemanni* Gyllenhal that is threatened in Finland and in Sweden (Rassi et al. 2000, Anon 2005). Also, little is known on the impact of fire on ecosystem processes (e.g. fire dependence) and the consequent long-term persistence of many forest species. Earlier Spence et al. (1999) underlined the importance of accommodating complex stand structures associated with gap-dynamic processes characteristic of post-rotational forests to conserve the biota and the processes that ensure the maintenance of site productivity.

One of the most pressing problems of applying smaller clear-cuts is the increasing amount of edges at the landscape level, as well as the total area that has to be managed to gain the same economic benefits as if clear-cuts were fewer in number but larger in size. Therefore, in the highly altered Fennoscandian forested landscape, restoration efforts should perhaps be focused on forests adjacent to existing reserves for the maximum benefit for threatened forest species, rather than spreading these activities evenly (and thinly) over large areas (Hanski 2000). However, in those parts of the North American and Russian boreal forest, where the history of intensive management is shorter and, thus, large-scale losses of forest specialist species are less likely yet to have taken place, the development and use of logging methods and landscape-scale ecological assessment of management strategies may still be a more efficient way to conserve forest diversity.

Needs for future research on carabids and forest management

This review has shown that most boreal forest carabids can survive in the forested landscape modified by forestry practices. However, the value of using carabids in forest research comes from the assumption that the management practices least affecting carabids are also those that have less of an impact on other forest taxa. Whether carabids truly act as indicators of the well-being of intact forests (and other habitats), including their most sensitive fauna and flora, should be evaluated. Among carabids there are sensitive species requiring old-growth conditions or other aspects of natural dynamics (such as fire) that are at risk in the managed forest landscape. Modern forestry practices aim at retaining certain key biotopes (such as riparian forest strips) and leaving more dead and live trees in the forest after harvest in order to enhance the survival of such old-growth species. However, the efficiency of these practices in maintaining forest species communities and especially in enhancing the survival of endangered species is currently poorly understood. Thus, the question to be answered by scientists remains: do these practices work, i.e. do they guarantee the survival of sensitive forest species?

Finally, we would like to point out a few other knowledge gaps on the relationship between carabids and forestry. First, the large-scale effects of landscape structure and processes on the distribution and abundance of boreal carabids have been little studied so far. Second, it is largely unknown how climate change will affect carabids and their biotopes in the near future, and whether these changes interact with the effects of forestry. Third, different large-scale management and conservation strategies should be experimentally evaluated. At a smaller scale, additional issues arise. First, certain forest types have been little studied; these include, for example, coastal North European pine forests. Second, we lack specific information on critical amounts of (micro-) habitat and forest characteristics, such as downed dead wood, on carabids. Third, despite the remarkable contributions of many great carabidologists—most notably Carl Lindroth—on species ecology, we still know very little on the dispersal abilities of carabid beetles, i.e. distances covered by foot and by flying, and how frequently carabids actually rely on these abilities and under which conditions. Also carabid behaviour in habitats other than their preferred ones is largely unknown. Clearly, we need to understand the importance of these aspects better than we currently do, as they affect carabid survival in the managed forested landscapes and, more generally, in the rapidly changing world.

Acknowledgments MK thanks the University of Alberta for funding (Izaak Walton Killam Postdoctoral Fellowship).

References

Abildsnes J, Tømmerås BÅ (2000) Impacts of experimental habitat fragmentation on ground beetles (Coleoptera, Carabidae) in a boreal spruce forest. Ann Zool Fenn 37:201–212

Andersson LI, Hytteborn H (1991) Bryophytes and decaying wood: a comparison between managed and natural forest. Holarct Ecol 14:121–130

Andrén H (1995) Effects of landscape composition on predation rates at habitat edges. In: Hansson L, Fahrig L, Merriam G (eds) Mosaic landscapes and ecological processes. London, Chapman & Hall, pp 225–255

Angelstam P (1997) Landscape analysis as a tool for the scientific management of biodiversity. Ecol Bull 46:140–170

Angelstam P (1998) Maintaining and restoring biodiversity by developing natural disturbance regimes in European boreal forest. J Veget Sci 9:593–602

Angelstam P, Pettersson B (1997) Principles of present Swedish forest biodiversity management. Ecol Bull 46:191–203

Anon (2005) Official Swedish red lists. SLU ArtDatabanken, Sweden; http://www.artdata.slu.se/rodlista (Accessed date 27 February 2006)

Atlegrim O, Sjöberg K, Ball JP (1997) Forestry effects on a boreal ground beetle community in spring: selective logging and clear-cutting compared. Entomol Fenn 8:19–26

Ås S (1993) Are habitat islands islands? Woodliving beetles (Coleoptera) in deciduous forest fragments in boreal forest. Ecography 16:219–228

Bauer LC (1989) Moorland beetle communities on limestone 'habitat islands'. I. Isolation, invasion and local species diversity in carabids and staphylinids. J Anim Ecol 58:1077–1098

Beaudry S, Duchesne LC, Cote B (1997) Short-term effects of three forestry practices on carabid assemblages in a jack pine forest. Can J For Res 27:2065–2071

Bonan GB, Shugart HH (1989) Environmental factors and ecological processes in boreal forests. Annu Rev Ecol Syst 20:1–28

Burke D, Goulet H (1998) Landscape and area effects on beetle assemblages in Ontario. Ecography 21:472–479

Chen J, Franklin JF, Spies TA (1995) Growing-season microclimatic gradients from clearcut edges into old-growth Douglas-fir forests. Ecol Appl 5:74–86

Davies KE, Margules CR (1998) Effects of habitat fragmentation on carabid beetles: experimental evidence. J Anim Ecol 67:460–471

de Warnaffe GDB, Lebrun P (2004) Effects of forest management on carabid beetles in Belgium: implications for biodiversity conservation. Biol Conserv 118:219–234

Debinski DM, Holt RD (2000) A survey and overview of habitat fragmentation experiments. Conserv Biol 14:342–355
den Boer PJ (1990a) Density limits and survival of local populations in 64 carabid species with different powers of dispersal. J Evol Ecol 3:19–48
den Boer PJ (1990b) The survival value of dispersal in terrestrial arthropods. Biol Conserv 54:175–192
Desender K, Ervynck A, Tack G (1999). Beetle diversity and historical ecology of woodlands in Flanders. Belg J Zool 129:139–155
Dettki H, Esseen P-A (1998) Epiphytic macrolichens in managed and natural forest landscapes: a comparison at two spatial scales. Ecography 21:613–624
Didham RK (1997) The influence of edge effects and forest fragmentation on leaf litter invertebrates in central Amazonia. In: Laurance WF, Bierregaard RO Jr (eds) Tropical forest remnants: ecology, management, and conservation of fragmented communities. Chicago & London, The University of Chicago Press, pp 55–70
Didham RK, Lawton JH (1999). Edge structure determines the magnitude of changes in microclimate and vegetation structure in tropical forest fragments. Biotropica 31:17–30
Donovan TM, Jones PW, Annand EM, Thompson FR III (1997) Variation in local-scale edge effects: mechanisms and landscape context. Ecology 78:2064–2075
Esseen P-A, Ehnström B, Ericson L, Sjöberg K (1997) Boreal forests. Ecol Bull 46:16–47
Ewers RM, Didham RK (2006) Confounding factors in the detection of species responses to habitat fragmentation. Biol Rev 81:117–142
Fahrig L, Merriam G (1994) Conservation of fragmented populations. Conserv Biol 8:50–59
Fournier E, Loreau M (2001) Respective roles of recent hedges and forest patch remnants in the maintenance of ground-beetles (Coleoptera: Carabidae) diversity in an agricultural landscape. Landscape Ecol 16:17–32
Fries C, Johansson O, Pettersson B, Simonsson P (1997) Silvicultural models to maintain and restore natural stand structures in Swedish boreal forests. For Ecol Manage 94:89–103
Gandhi KJK, Spence JR, Langor DW, Morgantini LE (2001) Fire residuals as habitat reserves for epigaeic beetles (Coleoptera: Carabidae and Staphylinidae). Biol Conserv 102:131–141
Greenslade PJM (1983) Adversity selection and the habitat templet. Am Nat 122:352–365
Haila Y (1999) Islands and fragments. In: Hunter ML Jr (ed) Maintaining biodiversity in forest ecosystems. Cambridge, Cambridge University Press, pp 234–264
Haila Y, Hanski IK, Niemelä J, Punttila P, Raivio S, Tukia H (1994) Forestry and the boreal fauna: matching management with natural forest dynamics. Ann Zool Fenn 31:187–202
Hallman E, Hokkanen M, Juntunen H, Korhonen KM, Raivio S, Savela O, Siitonen P, Tolonen A, Vainio M (1996) Alue-ekologinen suunnittelu. Vantaa, Finnish Forest and Park Service [In Finnish]
Halme E, Niemelä J (1993) Carabid beetles in fragments of coniferous forest. Ann Zool Fenn 30:17–30
Hanski I (1999) Metapopulation ecology. Oxford, Oxford University Press
Hanski I (2000) Extinction debt and species credit in boreal forests: modelling the consequences of different approaches to biodiversity conservation. Ann Zool Fenn 37:271–280
Heliölä J, Koivula M, Niemelä J (2001) Distribution of carabid beetles (Coleoptera, Carabidae) across boreal forest – clear-cut ecotone. Conserv Biol 15:370–377
Helle P, Muona J (1985) Invertebrate numbers in edges between clear-fellings and mature forests in Northern Finland. Silva Fenn 19:281–294
Heyborne WH, Miller JC, Parsons GL (2003) Ground-dwelling beetles and forest vegetation change over a 17-year-period, in western Oregon, USA. For Ecol Manage 179:123–134
Holliday NJ (1984) Carabid beetles (Coleoptera, Carabidae) from burned spruce forest (*Picea* spp.). Can Entomol 116:919–922
Holliday NJ (1991) Species responses of carabid beetles (Coleoptera, Carabidae) during post-fire regeneration of boreal forest. Can Entomol 123:1369–1389
Holliday NJ (1992) The carabid fauna (Coleoptera: Carabidae) during postfire regeneration of boreal forest: properties and dynamics of species assemblages. Can J Zool 70:440–452
Hunter ML (1996) Fundamentals of conservation biology. Blackwell Science Inc
Hänggi A, Baur B (1998) The effect of forest edge on ground-living arthropods in a remnant of unfertilized calcareous grassland in the Swiss Jura mountains. Mitt Schw Entomol Gesell Bul Soc Entomol Suisse 71:343–354
Imponen V, Kaila S (1988) Uudistusalan koon vaikutus metsän uudistamiskustannuksiin Etelä-Suomen oloissa. Metsätehon Katsaus 18:1–6 [In Finnish]
Jalonen J, Vanha-Majamaa I (2001) Immediate effects of four different felling methods on mature boreal spruce forest understorey vegetation in southern Finland. For Ecol Manage 146:25–34
Jokimäki J, Huhta E, Itämies J, Rahko P (1998) Distribution of arthropods in relation to forest patch size, edge, and stand characteristics. Can J For Res 28:1068–1072
Jonsell M, Weslien J, Ehnström B (1998) Substrate requirements of red-listed saproxylic invertebrates in Sweden. Biodivers Conserv 7:749–764
Karvonen L (2000) Guidelines for landscape ecological planning. Forestry Publications of Metsähallitus 36
Koivula M (2001) Carabid beetles (Coleoptera, Carabidae) in boreal managed forests – meso-scale ecological patterns in relation to modern forestry. PhD thesis, University of Helsinki, 120 pp
Koivula M (2002a) Alternative harvesting methods and boreal carabid beetles (Coleoptera, Carabidae). For Ecol Manage 167:103–121
Koivula M (2002b) Carabid beetles (Coleoptera, Carabidae) in thinned uneven-aged and clear-cut spruce stands. Ann Zool Fenn 39:131–149
Koivula M (2003) The forest road network – a landscape element affecting the spatial distribution of boreal carabid beetles (Coleoptera, Carabidae). In: Szyszko J et al (eds) How to protect or what we know about Carabid beetles. Warsaw Agricultural University Press, Warsaw, pp 287–299
Koivula M (2006) Effects of forest roads on spatial distribution of boreal Carabid beetles (Coleoptera: Carabidae). Coleopt Bull 59(4/2005):465–487
Koivula M, Niemelä J (2002) Boreal carabid beetles (Coleoptera, Carabidae) in managed spruce forests – a summary of Finnish case studies. Silva Fenn 36:423–436
Koivula M, Niemelä J (2003) Gap felling as a forest harvesting method in boreal forests: responses of carabid beetles (Coleoptera, Carabidae). Ecography 26:179–187
Koivula MJ, Vermeulen HJW (2005) Highways and forest fragmentation – effects on carabid beetles (Coleoptera, Carabidae). Landscape Ecol 20:911–926
Koivula M, Haila Y, Niemelä J, Punttila P (1999) Leaf litter and the small-scale distribution of carabid beetles (Coleoptera, Carabidae) in the boreal forest. Ecography 22:424–435

Koivula M, Kukkonen J, Niemelä J (2002) Carabid-beetle (Coleoptera, Carabidae) assemblages in the boreal forest succession. Biodivers Conserv 11:1269–1288

Koivula M, Hyyryläinen V, Soininen E (2004) Carabid beetles (Coleoptera: Carabidae) at forest-farmland edges in southern Finland. J Insect Conserv 8:297–309

Koivula M, Cobb T, Déchêne A, Jacobs J, Spence JR (2005) *Sericoda* Kirby, 1837 (Coleoptera: Carabidae) responses to time since fire, fire severity, post-fire logging, and smoke and charcoal stimuli. In: Serrano J, Gomez-Zurita J, Ruiz C (eds) XII European carabidologists meeting, Murcia, September 19–22, 2005. Ground beetles as a key group for biodiversity conservation studies in Europe. University of Murcia & David Mayntz, Murcia, pp 153–159

Kotze DJ, O'Hara RB (2003) Species decline – but why? Explanations of carabid beetle (Coleoptera, Carabidae) declines in Europe. Oecologia 135:138–148

Kotze DJ, Samways MJ (1999) Invertebrate conservation at the interface between the grassland matrix and natural Afromontane forest fragments. Biodivers Conserv 8:1339–1363

Kuusinen M (1994) Epiphytic lichen flora and diversity on *Populus tremula* in old-growth and managed forests of southern and middle boreal Finland. Ann Bot Fenn 31:245–260

Kuusinen M (1996) Importance of spruce swamp-forests for epiphyte diversity and flora on *Picea abies* in southern and middle boreal Finland. Ecography 19:41–51

Langor D, Spence J, Niemelä J, Carcamo H (1994) Insect biodiversity in the boreal forests of Alberta, Canada. In: Haila Y, Niemelä P, Kouki J (eds) Effects of management on the ecological diversity of boreal forests. Metsäntutkimuslaitoksen tiedonantoja 482. Finnish Forest Research Institute, pp 25–31

Laurance WF (2000) Do edge effects occur over large spatial scales? Trends Ecol Evol 15:134–135

Law BS, Dickman CR (1998) The use of habitat mosaics by terrestrial vertebrate fauna: implications for conservation and management. Biodivers Conserv 7:323–333

Liebherr JK, Song HJ (2002) Distinct ground beetle (Coleoptera: Carabidae) assemblages within a New York State wetland complex. J N Y Entomol Soc 110:127–141

Lindenmeyer DB, Franklin JF (2002) Conserving forest biodiversity. A comprehensive multiscaled approach. Washington DC, Island Press

Linder P, Östlund L (1992) Förändringar i norra Sveriges skogar 1870–1991 (Changes in the boreal forests of Sweden 1870–1991). Svensk Bot Tidskrift 86:199–215

Lindroth CH (1985) The Carabidae (Coleoptera) of Fennoscandia and Denmark. Fauna Entomol Scand 15:1–225

Lindroth CH (1986) The Carabidae (Coleoptera) of Fennoscandia and Denmark. Fauna Entomol Scand 15:233–497

Lindroth CH (1992) Ground beetles (Carabidae) of Fennoscandia. A zoogeographic study. Part I. Specific knowledge regarding the species. Washington D.C., Smithsonian Institution Libraries and the National Science Foundation. xxviii + 630 pp

Loope WL (1991) Interrelationships of fire history, land use history, and landscape pattern within Pictured Rocks National Lakeshore, Michigan. Can Field-Nat 105:18–28

Magura T, Kodobocz V, Tothmeresz B (2001) Effects of habitat fragmentation on carabids in forest patches. J Biogeogr 28:129–138

Magura T, Tothmeresz B, Elek Z (2005) Impacts of leaf-litter addition on carabids in a coniferous plantation. Biodivers Conserv 14:475–491

Martikainen P (2000) Effects of forest management on beetle diversity, with implications for species conservation and forest protection. PhD thesis, University of Joensuu

Martikainen P (2001) Conservation of threatened saproxylic beetles: significance of retained aspen *Populus tremula* on clearcut areas. Ecol Bull 49:205–218

Mascanzoni D, Wallin H (1986) The harmonic radar: a new method of tracing insects in the field. Ecol Entomol 11:387–390

Matlack GR, Litvaitis JA (1999) Forest edges. In: Hunter ML Jr (ed) Maintaining biodiversity in forest ecosystems. Cambridge, Cambridge University Press, pp 210–233

McCullough DG, Werner RA, Nuemann D (1998) Fire and insects in northern and boreal forest ecosystems of North America. Annu Rev Entomol 43:107–127

Moore JD, Ouimet R, Houle D, Camire C (2004) Effects of two silvicultural practices on ground beetles (Coleoptera: Carabidae) in a northern hardwood forest, Quebec, Canada. Can J For Res 34:959–968

Murcia C (1995) Edge effects in fragmented forests: implications for conservation. Trends Ecol Evol 10:58–62

Niemelä J (1993) Mystery of the missing species: species-abundance distribution of boreal ground beetles. Ann Zool Fenn 30:169–172

Niemelä J (1997) Invertebrates and boreal forest management. Conserv Biol 11:601–610

Niemelä J (1999) Management in relation to disturbance in the boreal forest. For Ecol Manage 115:127–134

Niemelä J (2001) Carabid beetles (Coleoptera, Carabidae) indicating habitat fragmentation: a review. Eur J Entomol 98:127–132

Niemelä J, Halme E (1992) Habitat associations of carabid beetles in fields and forests on the Åland Islands, SW Finland. Ecography 15:3–11

Niemelä J, Spence J (1999) Dynamics of local expansion by an introduced species: *Pterostichus melanarius* Ill. (Coleoptera, Carabidae) in Alberta, Canada. Divers Distrib 5:121–127

Niemelä J, Haila Y, Halme E, Pajunen T, Punttila P, Tukia H (1987) Habitat preferences and conservation status of *Agonum mannerheimii* Dej. in Häme, southern Finland. Notulae Entomol 67:175–179

Niemelä J, Haila Y, Halme E, Lahti T, Pajunen T, Punttila P (1988) The distribution of carabid beetles in fragments of old coniferous taiga and adjacent managed forest. Ann Zool Fenn 25:107–119

Niemelä J, Haila Y, Halme E, Pajunen T, Punttila P (1992) Small-scale heterogeneity in the spatial distribution of carabid beetles in the southern Finnish taiga. J Biogeogr 19:173–181

Niemelä J, Spence JR, Langor DW, Haila Y, Tukia H (1993a) Logging and boreal ground beetle assemblages on two continents: implications for conservation. In: Gaston KJ, New TR, Samways MJ (eds) Perspectives in insect conservation. Intercept Publishers Ltd, Andover, Hampshire, pp 29–50

Niemelä J, Langor DW, Spence JR (1993b) Effects of clear-cut harvesting on boreal ground beetle assemblages (Coleoptera: Carabidae) in western Canada. Conserv Biol 7:551–561

Niemelä J, Haila Y, Punttila P (1996) The importance of small-scale heterogeneity in boreal forests: variation in diversity in forest-floor invertebrates across the succession gradient. Ecography 19:352–368

Niwa CG, Peck RW (2002) Influence of prescribed fire on carabid beetle (Carabidae) and spider (Araneae) assemblages in forest litter in southwestern Oregon. Environ Entomol 31:785–796

Økland B (1994) Mycetophilidae (Diptera), an insect group vulnerable to forestry practices? A comparison of clearcut, managed and semi-natural spruce forests in southern Norway. Biodivers Conserv 3:68–85

Pajunen T, Haila Y, Halme E, Niemelä J, Punttila P (1995) Ground-dwelling spiders (Arachnida, Araneae) in fragments of old forest and surrounding managed forests in southern Finland. Ecography 18:62–72

Parviainen J, Seppänen P (1994) Metsien ekologinen kestävyys ja metsänkasvatusvaihtoehdot. Metsäntutkimuslaitoksen tiedonantoja 511. Finnish Forest Research Institute [In Finnish]

Pearce JL, Venier LA, McKee J, Pedlar J, McKenney D (2003) Influence of habitat and microhabitat on carabid (Coleoptera: Carabidae) assemblages in four stand types. Can Entomol 135:337–357

Peck R, Niwa CG (2004) Longer-term effects of selective thinning on carabid beetles and spiders in the Cascade Mountains of southern Oregon. Northwest Sci 78:267–277

Poole A, Gormally A, Skeffington MS (2003) The flora and carabid fauna of a mature and regenerating semi-natural oak woodland in south-east Ireland. For Ecol Manage 177:207–220

Punttila P, Haila Y, Pajunen T, Tukia H (1991) Colonisation of clearcut forests by ants in the southern Finnish taiga: a quantitative survey. Oikos 61:250–262

Punttila P, Haila Y, Niemelä J, Pajunen T (1994) Ant communities in fragments of old-growth taiga and managed surroundings. Ann Zool Fenn 31:131–144

Rassi P, Alanen A, Hakalisto S, Hanski I, Lehikoinen E, Ohenoja E, Siitonen J, Suvantola L, Vitikainen O, Väisänen R, Kanerva T, Mannerkoski I (2000) Suomen lajien uhanalaisuus 2000. Uhanalaisten lajien II seurantatyöryhmä. Finnish Ministry of Environment, 432 pp [In Finnish]

Riecken U, Raths U (1996) Use of radio telemetry for studying dispersal and habitat use of *Carabus coriaceus* L. Ann Zool Fenn 33:109–116

Saint-Germain M, Larrivée M, Drapeau P, Fahrig L, Buddle C (2005) Short-term response of ground beetles (Coleoptera: Carabidae) to fire and logging in a spruce-dominated boreal landscape. For Ecol Manage 212: 118–126.

Saunders DA, Hobbs RJ, Margules CR (1991) Biological consequences of ecosystem fragmentation: a review. Conserv Biol 5:18–32

Savolainen J (ed) (1997) Metsälaki perusteluineen. Edita, Helsinki [In Finnish]

Schneider RR (2002) Alternative futures. Alberta's boreal forest at the crossroads. The Federation of Alberta Naturalists and the Alberta Centre for Boreal Research, Edmonton, 152 pp

Siitonen J (2001) Forest management, coarse woody debris and saproxylic organisms: Fennoscandian boreal forests as an example. Ecol Bull 49:11–42

Siitonen J, Martikainen P (1994) Occurrence of rare and threatened insects living on decaying *Populus tremula*: a comparison between Finnish and Russian Karelia. Scand J For Res 9:185–191

Siitonen J, Saaristo L (2000) Habitat requirements and conservation of *Pytho kolwensis*, a beetle species of old-growth boreal forest. Biol Conserv 94:211–220

Sippola A-L, Siitonen J, Punttila P (2002) Beetle diversity in timberline forests: a comparison between old-growth and regenerating areas in Finnish Lapland. Ann Zool Fenn 39:69–86

Spence JR, Langor DW, Niemelä J, Cárcamo HA, Currie CR (1996) Northern forestry and carabids: the case for concern about old-growth species. Ann Zool Fenn 33:173–184

Spence JR, Buddle CM, Gandhi KJK, Langor DW, Volney WJA, Hammond HEJ, Pohl GR (1999) Invertebrate biodiversity, forestry and emulation of natural disturbance: a down-to-earth perspective. In: Meurisse RT, Ypsilantis WG, Seybold C (eds) Proceedings: Pacific northwest forest and rangeland soil organism symposium 461, pp 80–90

Stevens SM, Husband TP (1998) The influence of edge on small mammals: evidence from Brazilian Atlantic forest fragments. Biol Conserv 85:1–8

Taboada A, Kotze DJ, Salgado JM (2004) Carabid beetle occurrence at the edges of oak and beech forests in NW Spain. Eur J Entomol 101:555–563

Usher MB, Field JP, Bedford SE (1993) Biogeography and diversity of ground-dwelling arthropods in farm woodlands. Biodivers Lett 1:54–62

Vance CC, Nol E (2003) Temporal effects of selection logging on ground beetle communities in northern hardwood forests of eastern Canada. EcoScience 10:49–56

van Dijk TS (1996) The influence of environmental factors and food on life cycle, ageing and survival of some carabid beetles. Acta Jutland 71:11–24

Vanha-Majamaa I, Jalonen J (2001) Green tree retention in Fennoscandian forestry. Scand J For Res 3(Suppl.):79–90

Väisänen R (1995) Perhoset. In: Raivio S (ed) Talousmetsien luonnonsuojelu – yhteistutkimushankkeen väliraportti. Metsähallinnon luonnonsuojelujulkaisuja, Sarja A, 43, pp 103–108 [In Finnish]

Väisänen R, Biström O, Heliövaara K (1993) Sub-cortical Coleoptera in dead pines and spruces: is primeval species composition maintained in managed forests? Biodivers Conserv 2:95–113

Västilä S, Herrala-Ylinen H (1999) Silviculture. In: Sevola Y (ed) Finnish statistical yearbook of forestry. Finnish Forest Research Institute, pp 101–148

Wallin H (1986) Habitat choice of some field-inhabiting carabid beetles (Coleoptera: Carabidae) studied by recapture of marked individuals. Ecol Entomol 11:457–466

Werner SM, Raffa KF (2000) Effects of forest management practices on the diversity of ground-occurring beetles in mixed northern hardwood forests of the Great Lakes Region. For Ecol Manage 139:135–155

Wikars L-O (1992) Skogsbränder och insekter. Entomol Tidskrift 113:1–11 [In Swedish with English summary]

Wikars L-O (1995) Clear-cutting before burning prevent establishment of the fire-adapted *Agonum quadripunctatum* (Coleoptera: Carabidae). Ann Zool Fenn 32:375–384

Wikars L-O, Schimmel J (2001) Immediate effects of fire-severity on soil invertebrates in cut and uncut pine forests. For Ecol Manage 141:189–200

With KA, Gardner RH, Turner MG (1997) Landscape connectivity and population distributions in heterogeneous environments. Oikos 78:151–169

Work TT, Shorthouse DP, Spence JR, Volney WJA, Langor D (2004) Stand composition and structure of the boreal mixedwood and epigaeic arthropods of the Ecosystem Management Emulating Natural Disturbance (EMEND) landbase in northwest Alberta. Can J For Res 34:417–430

J Insect Conserv (2007) 11:19–27
DOI 10.1007/s10841-006-9015-z

BEETLE CONSERVATION

Conservation status of *Prodontria* (Coleoptera: Scarabaeidae) species in New Zealand

B. I. P. Barratt

Received: 27 January 2006 / Accepted: 23 March 2006 / Published online: 7 November 2006

Abstract The genus *Prodontria* (Scarabaeidae: Melolonthinae), which comprises 16 described species, is an endemic southern New Zealand genus occupying lowland and coastal habitats, inter-montane basins and alpine environments. Many of the species have a limited distribution and face potential threats from habitat change and predation. The only species to be formally protected is *P. lewisii* Broun, commonly known as the Cromwell Chafer, which is now restricted to a 81-ha reserve in Central Otago. One undescribed lowland species might be extinct as a result of habitat modification. The current status of *Prodontria* species, the known distribution of species and potential threats are discussed.

Keywords *Prodontria* · Scarabaeidae: Melolonthinae · Conservation · New Zealand

Introduction

From their introduction to the first major taxonomic revision of the Melolonthinae (Coleoptera: Scarabaeidae) it is clear that Given and Hoy (1952) undertook the work prompted more by the pest status of melolonthines in New Zealand (especially the major pasture pest, grass grub (*Costelytra zealandica* (White)) than they were by any consideration of conservation. In their revision the genus *Lewisiella* Broun was synonymised under the genus *Prodontria* Broun, and seven species of *Prodontria* were recognised. A key to species based on adults was presented and a description of the final instar of *P. lewisii* given. Two further species were described in 1960 (Given 1960), and a further species in 1964 (Given 1964). Emerson and Barratt (1997) described seven new species, and formally synonymised *P. bicolorata* under *P. modesta* on the basis of a molecular study (Emerson and Wallis 1994), bringing the total of species currently recognised to 16 (Table 1). A further undescribed species nominally "sp. Five Rivers" was found in the Southland Plains area, but is not thought to have been recollected since the 1960s.

Prodontria is essentially a southern New Zealand genus with species occupying lowland and coastal habitats, inter-montane basins and alpine environments (Table 1). Figure 1 shows the type localities, but not the complete known range for each species, which is given in Table 1. Many species have apparently quite restricted distributions, although further collecting will undoubtedly expand the known range of some species.

The adult stage of members of the genus characteristically feed at night, often on a wide range of herbaceous and woody shrub species, including exotic plant species. Based on observations of *Prodontria* species where studies have been undertaken (e.g. Ferreira et al. 1999) and the related genus *Odontria* (e.g. Barratt and Campbell 1982) the adults are active on nights when the temperature remains above a minimum threshold of about 7°C, particularly when humidity is above about 50% and with no more than light wind. During the day, and on nights when they are inactive, the beetles remain buried in the soil at depths of about 2–5 cm depending upon the substrate. However, some alpine species have frequently been

B. I. P. Barratt (✉)
AgResearch Invermay, Invermay Agricultural Centre, Puddle Alley, Private Bag 50034, Mosgiel, New Zealand
e-mail: barbara.barratt@agresearch.co.nz

Table 1 Described species of *Prodontria*, their type localities, known distribution and recognised threats

Prodontria species	Reference	Type locality	Distribution	Habitat	Locality code*	Recognised threats
capito	(Broun 1909)	Old Man Ra.	Central Otago mountains	Alpine herbfield	CO	Unknown
grandis	Given 1964	Mt. Anglem, Stewart Isl.	Mt. Anglem and Table Hill, Stewart Isl.	Alpine herbfield	SI	Rodents are potential predators, last recorded in 1991 (McGuinness 2001)
jenniferae	Emerson 1997	Southern Pisa Ra.	Southern Pisa Ra., Kawerau Gorge	Forest margins to subalpine	CO	Unknown
lewisii	Broun 1904	Cromwell on sandhills of Molyneux (now Clutha) River	Cromwell Chafer Nature Reserve (81 ha)	Intermontane basin, windblown sand dunes	CO	Predators, habitat modification (McGuinness 2001)
longitarsis	(Broun 1909)	The Snares Isl.	The Snares Isl.			Unknown
matagouriae	Emerson 1997	Sheldon Downs, Mackenzie Basin	Ben Ohau, Twizel, Mackenzie Basin, Burkes Pass	Intermontane basin shrubland		Apparently associated and possibly dependent upon *Discaria tomatou* (matagouri) (McGuinness 2001)
minuta	Emerson 1997	Tekapo River delta	Mackenzie Basin	Grassland		Habitat modification
modesta	(Broun 1909)	Manorburn, Central Otago	Alexandra area	Intermontane basin grassland	CO	Habitat modification from agricultural development, wilding pine and common thyme (McGuinness 2001)
montis	Emerson 1997	Rock and Pillar Ra.	Rock and Pillar Ra., Lammerlaw Ra.	Sub-alpine to alpine grassland/herbfield	CO	Unknown
patricki	Emerson 1997	Dansey's Pass	Dansey's Pass, Ida Ra., Omarama Saddle	Sub-alpine grassland		Unknown
pinguis	Given 1952	Hector Mts	Hector Mts., Mt Tennyson	Sub-alpine grassland	CO	Unknown
praelatella	(Broun 1919)	Near Lake Monowai	Coastal Southland and Stewart Isl.	Coastal dune vegetation	SL, SI	Unknown
rakiurensis	Emerson 1997	Table Hill, Stewart island	Table Hill, Mt Rakeahua	Sub-alpine to alpine grassland/herbfield	SI	Unknown
regalis	Emerson 1997	Crown Ra.	Crown Ra., Mt. Cardrona, Pisa Ra.	Alpine herbfield	CO	Unknown
setosa	Given 1952	Hunter Mts	Hunter Mts.	Unknown	FD	Unknown
truncata	Given 1960	Lake Monowai	Mt. Burns	Unknown	FD	Unknown
sp. 'Five Rivers'	–	–	Near Five Rivers, Southland Plains	Found in pasture	SD	May be extinct; habitat modification

*(Crosby et al. 1976)

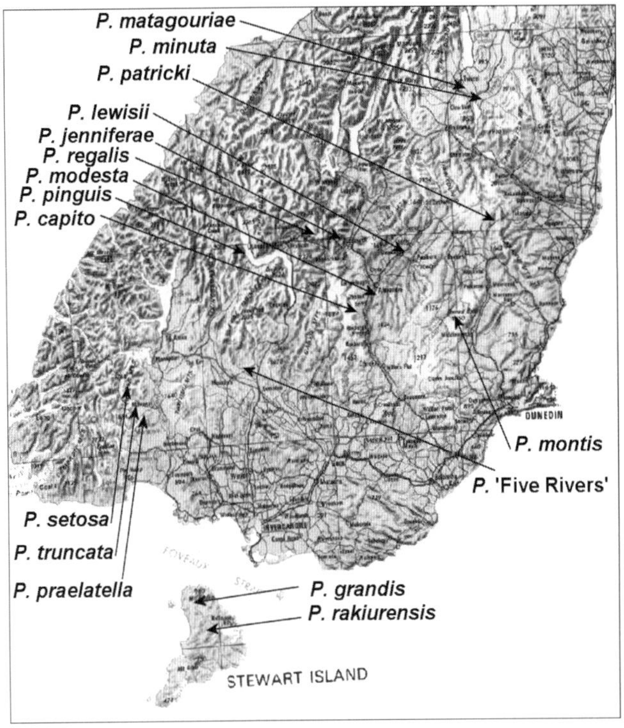

Fig. 1 Map of the southern South Island of New Zealand showing type localities for *Prodontria* species, excluding *P. longitarsis* (Snares Islands). *Prodontria* 'Five Rivers' shows the locality where this species was found

observed active by day, especially in calm, foggy conditions. The larvae are soil-dwelling root-feeders. They are thought to be most commonly associated with roots of grass species but these associations require further study.

Clarifying the conservation needs of *Prodontria* requires knowledge of the taxonomy of the genus, the biology and ecology of the species, and a detailed understanding of the distribution of the species. From this information, and knowledge of the threats present in its current range, the conservation status can be assessed and recovery plans formulated if required. In the case of *Prodontria*, this level of knowledge is incomplete for many species, but the objective of this contribution is to update, review and discuss the current state of knowledge of these factors.

Species of known conservation concern

Prodontria lewisii Broun

The most well recognised and researched species is *P. lewisii*, commonly known as the Cromwell Chafer. This species was described by Broun (1904) from three mutilated specimens collected by J.H. Lewis "on the sandhills of the Molyneux river" in 1903. Between then and 1968, when the Department of Scientific and Industrial Research scientist Charles Watt first visited Cromwell, only a few adults and three larvae had been collected. Because of its very specialised habitat requirements and limited range of distribution, Watt (1977) declared *P. lewisii* to be "in imminent danger of extinction". Extensive searching and pitfall trapping by Watt at apparently similar sites in Central Otago failed to reveal the Cromwell chafer anywhere other than in and near Cromwell (Watt 1979). A submission was made to the Cromwell Borough Council Joint Planning Committee to establish a reserve and to fence the proposed reserve area (Watt 1975). This was undertaken in 1979 and then in 1982 the Crown bought the land from the Council and added a small area of existing Crown land. This formed an 81-ha reserve that was administered by the then Department of Lands and Survey, now the Department of Conservation (DOC). The reserve was gazetted as the Cromwell Chafer Nature Reserve (CCNR) under the Reserves Act (1977) in 1983.

Watt (1979) stated that the beetle may never have occurred naturally beyond the 500 ha of the distinctive wind-blown Cromwell sandy loam dune system which the species inhabits, making it one of the smallest natural ranges for any known species. Recent forestry and vineyard development in the Cromwell area would almost certainly have precluded any significant populations from surviving beyond the immediate vicinity of the Nature Reserve.

Concern had been expressed that the population of silver tussock plants (*Poa cita*) in the reserve was declining, and that exotic grasses were becoming dominant. Roots of silver tussock were suggested by Watt (1979) to be the main food source for *P. lewisii* larvae. A vegetation survey carried out in 1990 showed little botanical diversity in the reserve, recording only 35 vascular plants of which only 14 were native. *P. cita* was present at <5% cover, and exotic species such as sheep's sorrel (*Rumex acetosella*), St. John's wort (*Hypericum perforatum*) and sweet vernal (*Anthoxanthum odoratum*) had become dominant in some parts of the reserve (Rance 1990).

Apart from changes in vegetation composition, recognised threats to *P. lewisii* are vertebrate predators. The castings of the little owl (*Athene noctua*) were found to comprise on average 10% *P. lewisii* remains, but the impact of this on population dynamics is unknown (Armstrong 1987). Hedgehogs were also found to be predators of *P. lewisii*, but again impact has not been determined. However, Brignall-Theyer (1998) considered hedgehogs to be more significant predators of *P. lewisii* than little owls. Rabbits have been abundant in the reserve at times, and although now controlled to low levels, they were responsible for considerable damage to the vegetation. Rabbit burrows damaged substrate structure particularly in the dune areas where the largest populations of *P. cita* were located. A proposal to site a refuse landfill area in the vicinity of the reserved raised concerns that increased rat populations might become another important source of predation. The landfill did proceed but extensive measures such as predator-proof fencing and rat control programmes were taken to avoid impacts on the reserve.

Considerable effort has gone in to mapping the distribution, and monitoring populations of *P. lewisii* in the CCNR (Armstrong 1987, 1990; Barratt and Patrick 1992; Emerson 1993; Ferreira 1997; Hunt 1998). Most of the research carried out has depended upon pitfall trapping and mark/recapture methods for population assessment (Ferreira 1997; Ferreira et al. 1999). Ferreira et al. (1999) concluded from their study that new sampling methods may be necessary to refine the conservation management plan for *P. lewisii* and that management will depend upon a better understanding of the habitat needs of the immature stages. The latter was also emphasised by Emerson (1994). Until recently, no attempt has been made to quantitatively assess larval population density, partly because of the intrinsically destructive nature of such an activity, and also because exploratory digging had shown that it was difficult to locate larvae in the soil (Barratt, unpublished). It had been suggested by Watt (1979), but unproven, that the immature stages feed on the roots of *P. cita*. It was therefore considered an important priority to determine the feeding requirements of *P. lewisii* larvae so that the CCNR can be managed appropriately to conserve the species. This, coupled with concerns expressed about possible declining numbers of adult *P. lewisii* in the reserve based upon data from consecutive seasons of pitfall trapping, prompted the initiation of a larval sampling programme to provide quantitative data on *P. lewisii* population fluctuations. Monitoring larval populations was considered likely to provide a more reliable indicator of emerging trends in population change, and hence potential for intervention than adult counts which are affected by climatic variables, and therefore not quantitative.

In 2001, a study to develop a long-term quantitative sampling strategy for *P. lewisii* was initiated. The objectives of the study were to obtain a general overview of the density and spatial distribution of *P. lewisii* immature stages in the entire CCNR in relation to vegetation and soil physical characteristics, and using this information, to develop a longer-term sampling strategy for the reserve. Initially the whole reserve was sampled by taking soil cores 9.3 cm in diameter and up to 35 cm deep (where the substrate allowed). Samples were taken 25 m apart along transect lines 25 m apart, resulting in a total of 1314 cores which were returned to the laboratory and hand-sorted. Disappointingly, this major exercise resulted in only two *P. lewisii* larvae being found, equating (tentatively) to an overall density in the reserve of about 1 larva per 5 m^2. Both larvae were found in a central location in the reserve, 75 m from each other, close to an area found during previous night search and mapping studies to have a relatively high concentration of *P. lewisii* adult activity.

As a result of the reserve-wide survey, an intensive core-sampling programme was initiated in the area of highest *P. lewisii* adult population density. The area was divided into 42 20 × 20 m plots and on each sampling occasion 16 cores (as described above) are taken per plot. The plots are sampled annually in December (excluded 2002) and sample processing is carried out on site so that scarab larvae (and adults) and all other invertebrates can be identified and returned to the soil. Results of sampling to date are shown in Fig. 2, which indicate population density in this area to be between 1 and 5 per m^2. A power law calculation was carried out using the variances from two sample dates. This showed that to detect a significant 50% change in *P. lewisii* density from one sampling date to the next with 90% confidence, core sample numbers would need to be increased to about 5,000. Clearly this would be unacceptably destructive even if it was practically possible, and so a compromise has been adopted where quantitative data are being collected which will at least allow comparisons to be made over time. Clearly this sampling programme needs to be continued for a

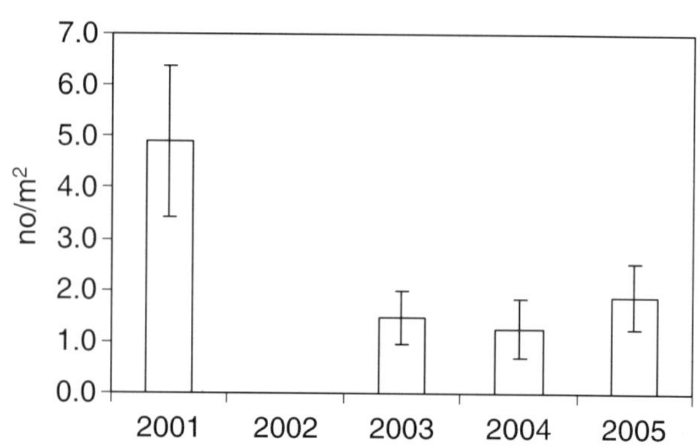

Fig. 2 Mean density (±s.e.) of *P. lewisii* larvae plus adults calculated from December core samples taken in the Cromwell Chafer Nature Reserve

number of years to detect any real trends in population density change, and to monitor effects of management changes.

Invertebrate populations in general in this relatively arid and poorly vegetated environment were found to be present at low densities. In conjunction with annual sampling, associated studies carried out in the reserve have attempted to correlate spatial distribution of *P. lewisii* larvae with substrate characteristics and vegetation, but no strong correlations have been detected. Plans for future research include an investigation of larval feeding preferences and growth rates to determine whether roots of silver tussock are indeed the optimum food source. An investigation of potential invertebrate predators (e.g. of eggs and larvae) and pathogens (e.g. microsporidia) will be undertaken to determine the sigificance of these, and potential for management.

Hopefully, some of the findings from the relatively intensive study of one species of *Prodontria* will assist in making predictions about the ecological requirements of other species in the genus. Naturally, such predictions should be treated with caution until they can be validated.

Prodontria modesta (Broun)

Prodontria modesta and *P. bicolorata* were described from the vicinity of Alexandra in Central Otago. Here horticulture, viticulture and airport development were considered to be risks to the grassland environment and hence survival of the species. Even though they were described as distinct species, in so doing, Given and Hoy (1952), and later Watt (1979) discussed the possibility that they might in fact be colour variants of the same species, since in terms of structural morphology, including male genitalia, they are indistinguishable. The colour morphs, however, are quite distinct, '*modesta*' being uniformly very dark in colour and '*bicolorata*' being pale brown with darker brown markings. The DOC were keen to resolve the issue of the integrity of these species since their limited distribution, particularly '*bicolorata*' was considered of conservation concern. Using night-searching during spring for two consecutive years, Emerson (1994) mapped the distribution of these species and found them to occur only within a radius of about 10 km around the town of Alexandra in Central Otago. In a small area of sympatry in the western part of their range the two morphs were observed interbreeding. Part of this area of sympatry near Conroy's Dam is now managed by DOC in the Aldinga Reserve.

Emerson and Wallis (1994) using isozyme analysis and principle component analysis of the data found that there were differences between populations of the two colour morphs within their area of distribution, but in the area of sympatry, the allelic frequencies were not significantly different at the 95% level. In concluding, the authors emphasised the importance for conservation of obtaining robust intraspecific data with which to compare between-species variability, since had two individuals from the extreme parts of the range been compared, they might have concluded that the two morphs were in fact distinct species. Using mtDNA COII sequence analysis of a number of species of *Prodontria* (Emerson and Wallis 1995) showed that the two morphs are sister taxa and differed by only 85%, the lowest value in the analysis, which provided further circumstantial evidence to validate the single species hypothesis (Wallis 2001). On the basis of these studies, *P. bicolorata* was synonymised with *P. modesta* (Emerson and Barratt 1997). Vink et al. (2003) confirmed the validity of this synonymy using phylogenetic analysis of a section of the CO1 gene region which showed no monophyletic lineages linked to colour form, and maximum inter-population variation of 2.25% which is far less than is known from other closely related beetle species.

Species of unknown conservation concern

Island-dwelling species

Prodontria species living on Stewart Island (*P. grandis, P. praelatella, P. rakiurensis*) and The Snares (*P. longitarsis*) are probably not at risk from habitat modification in the foreseeable future, but the potential risks from predators on these islands has been acknowledged (McGuinness 2001). The Snares islands are 209 km to the SW of the South Island of New Zealand and they are completely free of introduced land mammals. They are closed to the public and boat landings are by permit only. A survey of *P. longitarsis* was carried out on The Snares in 1987, but no further observations have been made since then. The habitat is intact and so the population is considered to be secure, and the major risk would be the remote possibility of rodent establishment (Brian Rance, DOC, pers. comm.).

The risk of predators such as rats, cats, and stoats to endangered bird species on New Zealand offshore islands is being addressed, and DOC have had considerable success in predator eradication. Clearly these activities are also likely to benefit the invertebrate fauna, including flightless melolonthines such as *Prodontria*.

Sub-alpine and alpine-dwelling species

Of the species living in subalpine and alpine environments, only *Prodontria capito* is known to have a relatively wide distribution. *P. capito* occurs on most of the Central Otago mountain ranges such as the Old Man Ra. (type locality), the adjacent Umbrella Mts., Old Woman Ra., Garvie Mts., the Remarkables and adjacent Hector Mts., Pisa Ra., Crown Ra. and Mt Cardrona. Some of these ranges have been partly developed as ski fields, an activity which can result in extensive habitat modification. Agricultural development is unlikely at these higher altitudes and so is unlikely to put this species at risk. Interestingly, the specimens of *P. capito* found on the Umbrella Mts. are mostly of a slightly different colour pattern to specimens found throughout the rest of its range, which are uniform black in colour. The Umbrella Ra. specimens are often dark brown with dark mottling on the elytra. Whether or not these specimens are genetically distinct is unknown, but otherwise morphologically they are indistinguishable from the black form.

A black species of *Prodontria* found on the Rock & Pillar and Lammermoor Ranges was until recently thought to be *P. capito*, but Emerson and Barratt (1997) noted a number of consistent distinguishing features including the shape of the aedeagus, and consequently the species was described as *P. montis*. The species is apparently restricted to these ranges, large parts of which are now managed by DOC and hence likely to be well protected except from events such as fire. Although escaped and accidental fires are not frequent in these ranges, they do occur in dry summers. However, the larval population in the soil is unlikely to be at risk.

Until recently, *P. pinguis* was thought to be very restricted in its distribution to the sub-alpine to alpine zones of a small section of the Hector Ra. in Central Otago (Fig. 1). However, this was a consequence of low collecting effort, and it has subsequently been found on the other side of the Nevis Valley at nearby Mt Tennyson (Brian Patrick, Otago Museum pers. comm.). Both these areas may be managed by the DOC in the future as a result of the high country tenure review process (implementation of the Crown Pastoral Land Act 1998) by which pastoral lease land which is identified as containing conservation values is retained by the Crown.

P. patricki, discovered by Brian Patrick on Dansey's Pass in 1979 was only known from a single specimen from the summit above the pass (1200 m) until more specimens were found by the author from pitfall trapping in 1986. Since then remains of specimens have been found on nearby Mt Buster (Ida Ra.) and live adults 70 km to the NW on the Omarama Saddle (Brian Patrick pers. comm.). While there appears to be no current risk to the habitats where this species is found, the type locality at Dansey's Pass is freehold land still being farmed.

P. regalis was also known for several years from relatively few specimens collected on the Crown Ra. in Central Otago (Fig. 1), but more recently specimens have also been collected on nearby Mt. Cardrona. Morphologically this species is very similar to *P. pinguis* and Emerson and Wallis (1995) considered them to be sister species as a result of their phylogenetic analysis. A road over the Crown Range has recently been sealed and this has undoubtedly increased the volume of traffic, and possibly the risk of more frequent summer grassland fires on the range, which could directly and indirectly (through vegetation change following fire) become a threat to *P. regalis*. Should further development of the Mt Cardrona ski field occur, this could also reduce the available habitat for the species.

While these sub-alpine and alpine species cannot be considered to be of serious conservation concern, many have limited distributions, and hence they are vulnerable to any disturbance factors which might bring about habitat degradation.

Lowland to upland-dwelling species

Two *Prodontria* species are known from the Mackenzie Basin area (inland North Otago – South Canterbury), which represents the northern-most part of the known range for the genus. Both species have been discovered with the last 40 years. *P. matagouriae* was first collected in 1968, and it was named after its association with the woody shrub matagouri (*Discaria toumatou*). Although larval feeding preferences are unknown, most adults have been found in soil around the roots of matagouri and observed feeding on the vegetation. Clearly more work is required to understand more about this association, but clearance of this shrub from farmland could present a considerable risk to the species. Being a particularly prickly shrub, which often produces impenetrable thickets, farmers are often inclined to remove plants to facilitate stock movement. The shrub is widespread, and in itself of no conservation concern, but the distribution of *P. matagouriae* is not well defined, although it has been found about 30 km further north at Burkes Pass (Brian Patrick, pers. comm.) in a Scenic Reserve, where the apparent host plant is secure. *P. minuta*, also known from the Mackenzie Basin area is the smallest known species in the genus, and little is known of its distribution or ecology. This species was described from the Tekapo River 'delta' (Fig. 1), from specimens found in 1993.

P. jenniferae is another species about which very little is known, and hence the potential risks and conservation needs of the species are not known. Specimens have been found in forest margin habitats and in grassland up to the subalpine zone from the Kawarau Gorge up to 750 m at the southern end of the Pisa Ra., again a very small area of known distribution, but this nocturnally active species can be locally quite common (Brian Patrick, pers. comm.).

The two species from the western part of the range for the genus, *P. setosa* and *P. truncata* are both known from a small area south of Lake Manapouri in eastern Fiordland (Fig. 1). *P. setosa* was collected from the Hunter Mts. in 1925, and DOC has no information about its current conservation status, but can identify no particular concerns. *P. truncata* was described from near Lake Monowai (about 200–300 m in altitude) from a specimen collected in 1956, but remains of specimens of this species have also been found nearby on Mt Burns at 1250–1370 m collected from under damp rocks near a creek (John Marris, Lincoln Museum, pers. comm.), and up to 1525 m on Mt Burns in 1970 (Grace Hall, Landcare Research, pers. comm.). Remains of a further three specimens were collected in 1992–1993 at 1500 m on Mt. Burns (Cody Fraser, Otago Museum, pers. comm.). So this species may have quite a wide altitudinal range.

Prodontria 'Five Rivers' is a species known from only from specimens collected in 1968 from pasture near the township of Five Rivers on the Southland Plains (Fig. 1). There are 12 specimens held in the New Zealand Arthropod Collection (Grace Hall, Landcare Research, pers. comm.), all collected from pasture, or soil beside a stream. A specimen labelled Mid Dome, which is near Five Rivers, was found at 366 m at the very lower slopes of this mountain, and was also recorded from pasture. Some specimens were reared from larvae collected from pasture. This species might be the only known member of the genus to have become extinct (McGuinness 2001). However, the demise of this species has by no means been confirmed, and no recent collecting has been carried out to locate a population. However, intensive agricultural development in this area might well be responsible for the loss of suitable habitat for the species, possibly copper tussock communities. These were dominated by *Chionochloa rubra* ssp. *cuprea*, which was widespread in Southland, especially following forest fires lit by early Polynesians, but later largely replaced by exotic pasture species sown by European farmers.

Conservation of *Prodontria*

Information on the distribution and possible conservation concerns regarding the genus *Prodontria* has improved over the last 5–10 years as a result of further collecting, traditional and novel taxonomic revision (e.g. Emerson and Wallis 1995), conservation assessment (e.g. McGuinness 2001), and ecological studies (e.g. Barratt et al. 2003). While the CCNR is the only example of specific protection for a species in the genus, several species are found either on islands, or sub-alpine/alpine areas where threats from environmental change is reduced. Some species are likely to be protected through the 'tenure review' process. This has in the past resulted in mainly high altitude grasslands being managed for conservation, but does not exclude low altitude areas that contain identified values from being protected. Habitat loss is well recognised as a major threat to invertebrate communities. *Prodontria* is essentially a grassland-dwelling genus and hence susceptible to changes brought about by agricultural development and consequent disturbance by cultivation, grazing, burning and intensification. Threats from habitat invasion by woody weeds such as wilding pine, gorse and wild thyme remain, as does the threat from exotic mammalian and avian predation. That noted, Emerson (1994) pointed out the apparent resilience of *Prodontria* to habitat change exemplified by *P. modesta* which is found in highly modified wild thyme-dominated vegetation near Alexandra. The adults feed on thyme (*Thymus vulgaris*) plants, but what the larvae are feeding on is unknown. However, Jamieson (1998) found that when *T. vulgaris* occupied 50% or more ground cover, beetles were not able to survive in the consolidated substrate caused by the dense root structure. She found also that wilding pines could represent a threat to parts of the known range of *P. modesta*.

The status of *Prodontria* 'Five Rivers' is in urgent need of investigation, beginning with an intensive survey of the Five Rivers area to determine whether any surviving populations remain. Since very little is known about the species, phenological and behavioural information from relatively well known species such as *P. lewisii* could be used to design such a survey. This could include live pitfall trapping from early spring to mid-summer at a series of sites which are least disturbed in the area, ideally coupled with night searches on mild, calm nights during spring. If this was repeated over 2–3 seasons without success, then there could be some degree of confidence in concluding that the species is extinct. However, if one or more populations was discovered, then a recovery plan would be an important requirement for this species. Information on the habitat requirements of the species would enable safeguards to be put in place to enhance the survival of the species.

Public pride and interest in the conservation of native flora and fauna has a long history in New Zealand. Although most of this passion has focussed on forest vegetation and native birds, conservation of indigenous grassland landscapes and biota has also become embraced by the public in recent years and sizeable tussock grassland reserves have been gazetted. One such example is the Te Papanui Conservation Park, a 21,000 ha area of essentially tall tussock grassland in eastern Otago. The official opening of this park attracted the interest of large numbers of the public, and the press. The establishment of the CCNR was very strongly supported by residents of Cromwell, and much generous assistance was given to Dr Charles Watt in the 1970s in delimiting surveys, and collection of beetles from urban areas for relocation to the area designated for the reserve. The team of mainly amateur entomologists who rediscovered *P. pinguis* on the Hector Range in 1981 was greeted by a television crew who had travelled to the relatively remote Nevis Valley to record the event.

Further research and monitoring is essential in order to better assess risk to *Prodontria* species, and to improve understanding of their ecological requirements and resilience to environmental change. This is taking place for *P. lewisii*, but currently not for any other species. Given the limited distribution of this unique genus in New Zealand, and particularly that of some of the individual species, such studies should be given serious consideration.

Acknowledgements Research on *P. lewisii* was funded by the Department of Conservation. Thanks to Brian Patrick (Otago Museum) and Marcus Simmons for reviewing an earlier draft of this manuscript and the former also for collection information; Eric Edwards and Brian Rance, (Department of Conservation, Southland Conservancy) for collection information; Bruce McKinlay, Mike Tubbs and staff (Department of Conservation, Otago Conservancy) for support and assistance with research in the Cromwell Chafer Reserve; Ricardo Palma (Te Papa), John Early (Auckland Institute and Museum), John Marris (Lincoln University Museum) and Grace Hall (Landcare Research) for label information from specimens in their respective collections; Colin Ferguson and many AgResearch Invermay Biocontrol and Biosecurity staff for assistance with research on *P. lewisii*.

References

Armstrong HE (1987) Ecology of the Cromwell chafer beetle, Unpublished MSc. thesis, Department of Zoology, University of Otago, Dunedin, New Zealand

Armstrong HE (1990) Reassessment of the Cromwell chafer beetle population. Report for Department of Conservation, Otago Conservancy, Dunedin, 10 pp

Barratt BIP, Campbell RA (1982) Biology of the striped chafer, *Odontria striata* (Coleoptera: Scarabaeidae) I. The adult, flight and ground surface activity, female reproductive maturation, and food plant selection. N Z J Zool 9:249–266

Barratt BIP, Ferguson CM, Logan RAS (2003) Monitoring larval density of the Cromwell chafer, *Prodontria lewisii* Broun (Coleoptera: Scarabaeidae): sampling 2003. Report for Department of Conservation, 13 pp

Barratt BIP, Patrick BM (1992) Conservation of the Cromwell chafer. In: Proceedings of the 41st Entomological Society of New Zealand Annual Conference, Central Institute of Technology, Heretaunga, pp 39–42. Entomological Society of New Zealand

Brignall-Theyer ME (1998) Potential vertebrate predators of the Cromwell chafer beetle, Prodontria lewisii, Unpublished MSc. thesis, University of Otago, Dunedin, New Zealand

Broun T (1904) Description of new genera and species of New Zealand Coleoptera. Ann Mag Nat Hist 7:41–59

Crosby TK, Dugdale JS, Watt JC (1976) Recording specimen localities in New Zealand: an arbitrary system of areas and codes defined. N Z J Zool 3: 69 pp + map

Emerson BC (1993) Population monitoring of the Cromwell chafer beetle (*Prodontria lewisi*). Report for Department of Conservation, Otago Conservancy Dunedin, 22 pp

Emerson BC (1994) Conservation status of chafer beetles *Prodontria bicolorata* and P. modesta: distribution and ecological observations. N Z Entomol 17:3–6

Emerson BC, Barratt BIP (1997) Descriptions of seven new species of the genus *Prodontria* (Coleoptera: Scarabaeidae: Melolonthinae). Coleop Bull 53:23–36

Emerson BC, Wallis GP (1994) Species status and population genetic structure of the flightless chafer beetles *Prodontria modesta* and *P. bicolorata* (Coleoptera: Scarabaeidae) from the South Island, New Zealand. Mol Ecol 3:339–345

Emerson BC, Wallis GP (1995) Phylogenetic relationships of the *Prodontria* (Coleoptera; Scarabaeidae; subfamily Melolonthinae), derived from sequence variation in the mitochondrial cytochrome oxidase II gene. Mol Phylogenet Evol 4:433–447

Ferreira SM (1997). Conservation of the Cromwell chafer beetle (*Prodontria lewisi*): report on data and monitoring of population variables 1986–1997. Report for Department of Conservation, Dunedin, 35 pp

Ferreira SM, Ravenscroft NOM, McKinlay B (1999) Activity patterns and population characteristics of the New Zealand endemic Cromwell chafer, *Prodontria lewisii* (Scarabaeidae; Melolonthinae). N Z J Zool 26:229–241

Given BB (1960) The Melolonthinae (Coleoptera) of New Zealand: descriptions of new species, notes on types, and a catalogue. N Z J Sci 3:363–381

Given BB (1964) A new species of the genus *Prodontria* (Melolonthinae: Coleoptera). N Z J Sci 7:38–40

Given BB, Hoy JM (1952) A revision of the Melolonthinae of New Zealand. Report for New Zealand Department of Scientific and Industrial Research, Nelson, 172 pp

Hunt D (1998) Monitoring of the Cromwell Chafer Beetle (*Prodontria lewisii*) by pitfall trapping and night searches. Report for Department of Conservation, Dunedin, 32 pp

Jamieson CD (1998) Conservation management of *Prodontria* species in the Alexandra area. Report for Department of Conservation, Dunedin, pp 32

McGuinness CA (2001) The conservation requirements of New Zealand's nationally threatened invertebrates. Report for Department of Conservation, Wellington, 657 pp

Rance BD (1990) Cromwell chafer reserve scientific monitoring and vegetation. Report for Department of Conservation, Dunedin, 11 pp + Appendix

Vink CJ, Evans AM, Phillips CB, Murdoch TC, Tubbs MB (2003) Molecular phylogenetic analysis supports the synonymy of *Prodontria modesta* (Broun) and *Prodontria bicolorata* Given (Coleoptera: Scarabaeidae: Melolonthinae). J Insect Conserv 7:215 – 221

Wallis GP (2001) Taxonomic status of the chafer beetles *Prodontria modesta* and *P. bicolorata*. Report for Department of Conservation, Wellington, 13 pp

Watt JC (1975) Proposed biological reserve for the Cromwell chafer beetle, *Prodontria lewisii*. Report for Cromwell Joint Planning Committee and Cromwell Borough Council, 8 pp, 3 maps

Watt JC (1977) Conservation and type localities of New Zealand Coleoptera, and notes on collectors 1770–1920. J R Soc N Z 7:79–91

Watt JC (1979) Conservation of the Cromwell chafer, *Prodontria lewisii*(Coleoptera: Scarabaeidae). N Z J Ecol 2: 22–29

J Insect Conserv (2007) 11:29–30
DOI 10.1007/s10841-007-9067-8

ERRATUM

Conservation status of *Prodontria* (Coleoptera: Scarabaeidae) species in New Zealand

B. I. P. Barratt

Published online: 23 January 2007

Erratum to: J Insect Conserv DOI 10.1007/s10841-006-9015-z

Due to an unfortunate error, an incorrect version of Fig. 1 and Table 1 has been used in the above mentioned publication. On the following pages, the correct reproductions have been published and should be treated as definitive by the reader.

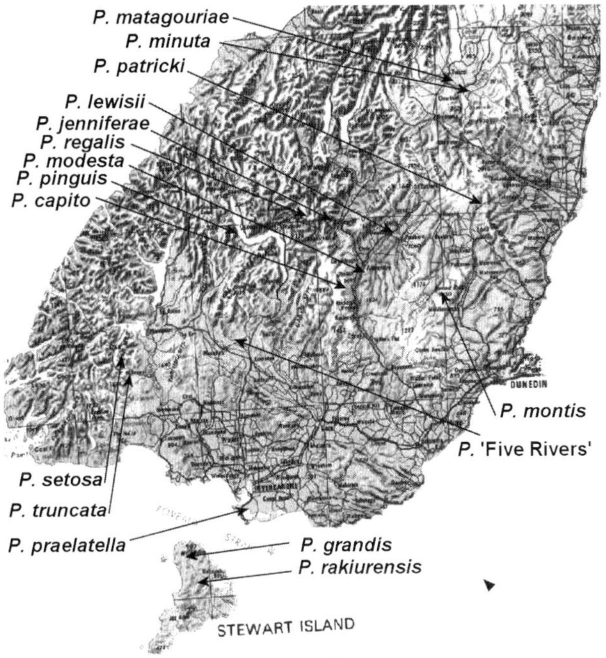

Fig. 1 Map of the southern South Island of New Zealand showing type localities for *Prodontria* species, excluding *P. longitarsis* (Snares Islands). *Prodontria* 'Five Rivers' shows the locality where this species was found

The online version of the original article can be found at http://dx.doi.org/10.1007/s10841-006-9015-z

B. I. P. Barratt (✉)
AgResearch Invermay, Invermay Agricultural Centre, Puddle Alley, Private Bag 50034, Mosgiel, New Zealand
e-mail: barbara.barratt@agresearch.co.nz

Table 1 Described species of *Prodontria*, their type localities, known distribution and recognised threats

Prodontria species	Reference	Type locality	Distribution	Habitat	Locality code*	Recognised threats
capito	(Broun 1909)	Old Man Ra.	Central Otago mountains	Alpine herbfield	CO	Unknown
grandis	Given 1964	Mt. Anglem, Stewart Isl.	Mt. Anglem and Table Hill, Stewart Isl.	Alpine herbfield	SI	Rodents are potential predators, last recorded in 1991 (McGuinness 2001)
jenniferae	Emerson 1997	Southern Pisa Ra.	Southern Pisa Ra., Kawerau Gorge	Forest margins to subalpine	CO	Unknown
lewisii	Broun 1904	Cromwell on sandhills of Molyneux (now Clutha) River	Cromwell Chafer Nature Reserve (81 ha)	Intermontane basin, windblown sand dunes	CO	Predators, habitat modification (McGuinness 2001)
longitarsis	(Broun 1909)	The Snares Isl.	The Snares Isl.		Snares	Unknown
matagouriae	Emerson 1997	Sheldon Downs, Mackenzie Basin	Ben Ohau, Twizel, Mackenzie Basin, Burkes Pass	Intermontane basin shrubland	MK	Apparently associated and possibly dependent upon *Discaria toumatou* (matagouri) (McGuinness 2001)
minuta	Emerson 1997	Tekapo River delta	Mackenzie Basin	Grassland	MK	Habitat modification
modesta	(Broun 1909)	Manorburn, Central Otago	Alexandra area	Intermontane basin grassland	CO	Habitat modification from agricultural development, wilding pine and common thyme (McGuinness 2001)
montis	Emerson 1997	Rock and Pillar Ra.	Rock and Pillar Ra., Lammerlaw Ra.	Sub-alpine to alpine grassland/herbfield	CO	Unknown
patricki	Emerson 1997	Dansey's Pass	Dansey's Pass, Ida Ra., Omarama Saddle	Sub-alpine grassland	CO	Unknown
pinguis	Given 1952	Hector Mts	Hector Mts., Mt Tennyson	Sub-alpine grassland	CO	Unknown
praelatella	(Broun 1909)	Invercargill (Greenhills)	Coastal Southland and Stewart Isl.	Coastal dune vegetation	SL, SI	Unknown
rakiurensis	Emerson 1997	Table Hill, Stewart island	Table Hill, Mt Rakeahua	Sub-alpine to alpine grassland/herbfield	SI	Unknown
regalis	Emerson 1997	Crown Ra.	Crown Ra., Mt. Cardrona, Pisa Ra.	Alpine herbfield	CO	Unknown
setosa	Given 1952	Hunter Mts	Hunter Mts.	Unknown	FD	Unknown
truncata	Given 1960	Lake Monowai	Mt. Burns	Unknown	FD	Unknown
sp. 'Five Rivers'	–	–	Near Five Rivers, Southland Plains	Found in pasture	SL	May be extinct; habitat modification

*(Crosby et al. 1976)

J Insect Conserv (2007) 11:31–41
DOI 10.1007/s10841-006-9016-y

BEETLE CONSERVATION

Carabid beetle (Coleoptera: Carabidae) conservation in New Zealand

C. A. McGuinness

Received: 31 January 2006 / Accepted: 20 April 2006 / Published online: 3 November 2006

Abstract An overview of carabid beetles in the New Zealand conservation context is provided. Processes for threatened species recovery within the New Zealand Department of Conservation are outlined, and the Department's two major beetle-related recovery documents (The Conservation Requirements for New Zealand's Nationally Threatened Invertebrates, and The Carabid Beetle Recovery Plan) are discussed. A subjective consideration of the merits and drawbacks of both the documents, and the systems that support them, is presented, along with an option for the way forward for threatened species conservation in New Zealand.

Keywords *Mecodema* · *Megadromus* · Predators · Recovery plans · Introduced vertebrates

Introduction

New Zealand lies in the Pacific Ocean, between 34°S and 47°S latitude, and covers an area of 270,534 square kilometres (Anon 2000). Its nearest sizeable neighbour is Australia approximately 1,930 km to the west. It has been geographically isolated from other landmasses for 80 million years (Cooper and Millener 1993), and many of New Zealand's invertebrate species are of Gondwanan origin (Klimaszewski and Watt 1997). This isolation, along with the changing climate, changing shorelines, orogonies, glaciation and volcanic activity, has enabled a highly distinctive biota to evolve (Klimaszewski and Watt 1997).

Species evolved during this long period of isolation with only one type of mammalian predator present —bats (*Mystacina* spp.). Many of New Zealand's native invertebrate species are ill equipped to deal with introduced mammalian predators. These native species often lack the morphological and behavioural adaptations required to successfully avoid mammalian predation. Many endemic invertebrates species developed gigantism and flightlessness, for example giant weevils of the genus *Lyperobius*. Many are nocturnal and use scent to communicate, some with odours easily detectable by humans (e.g. giant weta *Deinacrida* spp) (Field and Jarman 2001). These species generally rely on camouflage and remaining still to avoid detection by predators. Whilst this behaviour may be successful when dealing with many native predators, which are often visual hunters (e.g. tuatara, tomtit), it has proven to be woefully inadequate when dealing with nocturnal olfactory hunters.

Following the arrival of humans, 11 introduced mammals that prey on invertebrates have now naturalised in New Zealand. These include three species of rat (kiore, *Rattus exulans*; ship rat, *Rattus rattus*; Norway rat, *Rattus norvegicus*), the house mouse (*Mus musculus*), European hedgehog (*Erinaceus europaeus*), and three species of mustelid (stoats, *Mustela erminea*; ferrets, *Mustela furo*; and weasels *Mustela nivalis*). Feral pigs (*Sus scrofa*), whilst not usually considered predators, will prey on large invertebrates, in particular

C. A. McGuinness
Threatened Species Section, Department of Conservation, Wellington, New Zealand

Present Address:
C. A. McGuinness (✉)
Biosecurity New Zealand, Ministry of Agriculture and Forestry, Wellington, New Zealand
e-mail: carl.mcguinness@maf.govt.nz

giant landsnails (*Powelliphanta* spp) (Walker 2003a), and cause problems both through direct predation and habitat destruction. The introduced brushtail possum (*Trichosurus vulpecula*), and a number of introduced birds, such as the song thrush (*Turdus philomelos*), also have impacts on New Zealand's endemic invertebrates, particularly on some species of snail (e.g. *Powelliphanta* spp.).

Endemism and numbers of carabids

Levels of endemism for species assemblages in New Zealand are commonly around 80–90%. More than 5,000 native species of beetle are present, and over 340 introduced beetle species (Klimaszewski and Watt 1997). Ninety-two percent of the 424 species of Carabidae that have been recognised in New Zealand are endemic, and include representatives of 78 genera, 20 tribes and 5 subfamilies. The entire New Zealand carabid fauna may comprise about 600 species (Larochelle and Lariviere 2001).

Beetle conservation in New Zealand

Numbers of threatened beetle taxa

The New Zealand Department of Conservation (DOC) has established a threat classification system to assist with prioritising work on threatened species (Hitchmough and Bull 2005). Currently there are 28 "Acutely Threatened" beetle taxa (i.e. listed as either "Nationally Critical", "Nationally Endangered", or "Nationally Vulnerable"). Twelve of these are carabids. There are 131 "Chronically Threatened" beetle taxa (i.e. listed as either "Serious Decline", "Gradual Decline", "Range Restricted" or "Sparse"), with 33 being carabids. Add to this another 42 beetle taxa listed as "Data Deficient" (including three carabids), and there are 201 beetle taxa (48 carabids) which are listed as either threatened, or possibly threatened.

In relation to carabids, this is approximately 8% of their total predicted number. However, the majority of these carabids come from two genera—*Mecodema* (20 taxa comprising 17 valid and three tag-named taxa) and *Megadromus* (12 taxa comprising four valid and eight tag-named taxa). There are currently 64 *Mecodema* species recognised in New Zealand, and 24 *Megadromus* species (Larochelle and Lariviere 2001). Based on these numbers, and erring on the side of caution by excluding the tag-named taxa, approximately 31% of *Mecodema* species and 17% of *Megadromus*, are listed as threatened.

Threats to carabid beetles in New Zealand

As inferred earlier, the isolated evolution of New Zealand carabids has left them susceptible to predation from introduced mammalian predators. The large flightless carabids present in New Zealand, particularly in *Mecodema* and *Megadromus*, are believed to be most at risk of predation, although no rigorous data exist on population numbers or rates of decline. They are considered most at risk due to the fact that they are generally the larger carabid species, and are therefore believed to have fewer refuge sites available to them. Predation is believed to be the most significant threat currently facing these larger carabid species.

Habitat modification and loss have played a significant part in the decline in carabid range and numbers. Following the arrival of European settlers, vast tracts of bush, scrub and tussock grasslands were converted to pastoral grasslands and arable fields. Most of this occurred over the past 150 years. Most New Zealand carabids occur in vulnerable forest or tussock grassland habitats (Larochelle and Lariviere 2001). Some species of carabid are still suffering due to the large-scale forest clearance, particularly through the loss of refugia —namely, the large fallen logs that provide shelter and habitat. Introduced mammalian browsers, such as deer and goats, contribute to the modification of remaining forest habitat, opening up the understorey in forest systems and exposing it to more wind and desiccation. Predators may also impact on the habitat; for example, possums and rats eat seeds and seedlings, and so hinder regeneration. The true extent of loss to carabid species, and indeed invertebrates in general, caused by this modification or loss of habitat will probably never be known. Habitat loss and modification is still occurring in New Zealand, although not to the extent that it has previously.

The challenge facing conservationists in New Zealand is how to address these threats to assure the well-being of beetles and other taxa. The following sections will outline the process used to take a threatened species from being a name on a list, to an active management programme, using carabids to exemplify the various stages.

The Department of Conservation threatened species recovery documents

The Department utilises two main types of document to promote threatened species recovery—these are "Recovery Plans" and documents generally referred to as "Action Plans".

Action plans

The Department currently has three documents that can be considered as threatened species action plans—one on vascular plants, one on seabirds, and one on invertebrates.

Whilst no formal definition exists within the Department for "action plans", they do conform to a general format, and provide information on:

- Species distribution
- Threats
- Future management needs
- Research, survey or monitoring needs.

They are less detailed than recovery plans in regards to the management actions listed in them, and have no line accountability to enforce implementation (which recovery plans do have). Therefore they are less powerful documents in terms of leveraging action within the Department, but provide general guidance for anyone wanting to progress work on those species. They can also be useful advocacy tools, helping to raise awareness of the plight of the groups of species they cover.

Invertebrate specific management documents: the conservation requirements of New Zealand nationally threatened invertebrates—action plan

Background

In 1994 the Department listed 280 invertebrate taxa as national priorities for conservation (Molloy and Davis 1994). However, there was no centralised information available on these species and this proved a hindrance to management.

In 1998, a project was initiated to collate information on these taxa. The target audience was primarily Departmental staff, but also included research institutes. Their collective information needs were scoped, and the requisite information accumulated, leading to publication in 2001.

Purpose

The purpose of this document was to provide information on the key conservation requirements of threatened invertebrates, to assist the Department of Conservation staff to set national priorities for threatened invertebrate recovery programmes, and to establish key recovery actions.

Deliverables

The project culminated in the publication of a 658 page document—"The Conservation Requirements of New Zealand's Nationally Threatened Invertebrates" (McGuinness 2001). This provided information to assist with management and research relating to the species covered, as well as serving as a "pseudo field-guide" for the lay-person. The document gave an overview of invertebrates in New Zealand along with individual "profiles" of each of the taxa. These "profiles" covered:

- Taxonomic information
- A basic "lay-person" description
- Distribution
- Habitat
- Signs of presence
- Threats
- Work undertaken to date
- Priority research, survey and monitoring
- Management needs
- Contacts
- A photo and scale-line to aid identification.

Whilst collating this information, more than 500 additional taxa were mentioned as being of potential concern for conservation. These were listed in appendices and later ranked through the Department's threat ranking system.

Good points

The action plan collated all relevant information in one readily accessible place. It provided guidance on what the main threats were to species, and the high-level (non-detailed) management actions required. It also helped raise the profile of invertebrate conservation, through both the distribution of the document, and the associated media interest it generated.

Issues

The major drawback of this document, and other similar "action plans" in New Zealand is that they carry no line-accountability within the Department —they are a voluntary guide only, a resource that can be used if, and when, people wish. These plans remain as recommendations only, and no mechanism is present within the Department to enforce their implementation.

Recovery plans

Numbers

The Department of Conservation currently has 56 approved recovery plans (only five on invertebrates, the remainder being 29 on birds, three on fish, five on reptiles and amphibians, 12 on plants, one on bats, and one on marine mammals.). Collectively they cover over 100 taxa. The invertebrate plans cover giant land snails (*Placostylus* spp., *Paryphanta* spp. and *Rhytidarex* sp.) (Parrish et al. 2005), *Powelliphanta* land snails (Walker 2003a), weta (Sherley 1998), short-horned grasshoppers (Walker 2003b), and carabid beetles (McGuinness 2002).

Purpose

These plans are statements of the Department's intentions for the conservation of particular plants and animals for a defined period, usually 5–10 years. In focussing on goals and objectives for management, recovery plans serve to guide the Department in its allocation of resources and to promote discussion amongst a wider section of the interested public.

Implementation

Recovery plans are approved management documents with requirements for annual reporting. They operate under a direct line-accountability system, which means that in a hierarchical management structure such as that operated by the Department, a senior manager signs the document off as sponsor. Staff are then expected to implement the actions recommended in the plan, or explain why they were not implemented.

Once plans have been approved, the actions listed can then be incorporated into individual work-plans. The responsibility for undertaking a task outlined in the plan is assigned to the particular part of the organisation best suited to delivering it. For operational tasks, such as on-the-ground management, this is usually done at the Conservancy Office or Area Office level (a Conservancy is a region comprising a number of Area Offices. An Area Office covers a set geographic area. The size of these areas varies). Whilst recovery plans have line-accountability, and their implementation is expected, it is not mandatory —actions listed in the plan need to stack up as a priority alongside all the other work demands. The various tasks will be assessed at the level they are assigned. Assessment usually involves ranking in priority order, the various tasks that are competing for resources (both staff time and dollars).

Carabid beetle specific management documents: the carabid beetle recovery plan

The Carabid Beetle Recovery Plan (2002–2007) (McGuinness 2002) is a multi-species document covering 56 taxa. Rather than a single recovery plan, it is actually a series of recovery and action plans.

Structure of the document

The document begins with an introduction, followed by outlining the threats, ecology and biology, and main issues facing carabids in New Zealand. This is followed by the main body of the document—the species profiles and workplans.

Knowledge of the conservation status of the carabid taxa being covered by this document was highly variable. For some there was considerable information on distribution and threats, while others were only known from a handful of specimens from one, or a few, sites. The conservation actions appropriate for them therefore ranged from active management programmes through to habitat maintenance, opportunistic surveys, monitoring, or research. Due to this, the section covering species profiles and workplans was broken into four parts, grouping the taxa into areas based on the main type of work required. The parts were:

A. *Beetles with recovery plans*: Taxa requiring active management. Sufficient information was available on these beetles to believe they were declining and that there was some urgency to act. Most importantly there was confidence that the threats that had been identified could be managed. Four taxa came under this part of the document.
B. *Beetles requiring survey or other information gathering*: These taxa had insufficient information at hand to indicate decline. Survey and monitoring was required to determine the distribution and/or decline of the taxon. The taxa were grouped as high, medium or low priorities, based on current knowledge and perceived urgency for work. Twenty-four taxa were listed in this part of the document.
C. *Beetles requiring taxonomic clarification or an identification methodology*: For these taxa there was a need to be sure they were valid species before initiating recovery action. There was no great urgency to act associated with these taxa. Eleven taxa were listed in this part of the document.

D. Beetles previously listed as being of concern for which priority is currently downgraded and no action proposed at present. Seventeen taxa were listed in this part of the document along with the reasons for their exclusion from the plan.

Research needs to improve the basis for conservation of carabids as a whole were also considered in the document. The main areas of research that were determined as required are:

- Determination of a method for surveying and estimating population abundance, including the possible use of chemical attractants
- Taxonomic revision of some genera with priority given to *Mecodema* and *Megadromus*.

Beetles with recovery plans

Specific recovery actions were provided for four taxa (*Zecillenus tillyardi* (Brookes), *Mecodema costellum costellum* Broun, *Mecodema laeviceps* Broun, *Megadromus* "Eastern Sounds"). There was still little scientific evidence to back-up the concerns of decline for these taxa. However, these species are being acted on now, prior to conclusive evidence of decline. To wait until solid evidence of decline exists may be too late to save the species.

Species profiles and workplans were produced providing information on the following:

- Conservancy Office
- Area Office
- Past/present distribution
- Species ecology and biology
- Cause of decline and threats
- Past conservation efforts
- Issues
- Long term recovery goal
- Options for recovery
- Objectives for term of plan
- Workplan outlining:
    - Objectives
    - "SMART" performance measures (Specific, Measurable, Achievable, Realistic, Timebound)
    - Actions required
    - Priority
    - Responsibility

Note: taxa listed as requiring either "survey or other information gathering", or "taxonomic clarification or an identification methodology" also had profiles and action plans prepared for them. These contained similar information to the recovery plans, but excluded the long-term recovery goal, options for recovery, objectives for term of plan, and the "SMART" performance measures.

The workplans are the key management tool in this document. Actions associated with the workplans usually involve "research by management". The best available information is used to determine the threats and management action, then management is initiated along with monitoring and the gathering of information as the programme progresses. This allows determination of whether the key agent of decline has been identified, and whether the management action is effective or not. Management can be modified to make it more efficient, or, if deemed ineffective, stopped at any time and a new regime implemented.

Information gathering versus action

The primary aim of the recovery plans was to act now rather than await all information on biology, threats, and control methods, whilst the species continues to decline. If operating under the precautionary principle, as DOC does, then the prerogative is to act when available, albeit incomplete, information indicates a problem.

To act or not is a calculated risk—act now and possibly be wrong, or act later when the probability of being wrong is considerably less. Which is the correct option depends on the urgency and severity of the threat, and the scale of any potential negative impact on the species if the wrong decision is taken. Ultimately, the guiding question that needs be answered is—can the species afford to wait?

Taking positive action is highly unlikely to be worse than doing nothing; it will be mitigating some threat. There is always the possibility that any proposed management action will decrease the chances of survival of the species. For example, controlling a particular threat may exacerbate a previously secondary, low-level threat, by relieving the pressure that was on it. In New Zealand we face this risk when controlling rats, as the mouse population may explode. This secondary threat may then assert itself to be an even greater threat than the initially identified primary threat. Monitoring would normally detect this, allowing the management regime to be adapted accordingly. The key is to retain flexibility and openness to alternative approaches.

Only four carabids were deemed to be sufficiently well-known to enable management to be undertaken, with some confidence that the action would result in improved security of the taxa. For these species there was at least an educated guess available as to the agent of decline. No scientific study had been conducted to ascertain this, but management can start and the populations be monitored to see if the action is benefiting the species. If population numbers remain the same or decrease, then it is likely that the agent being managed is not the cause of the decline. Management can then be modified to target a different area of concern.

For example, with *Mecodema costellum costellum* found on Stephens Island, it was believed that the lack of fallen logs as refuge may be limiting the beetle's ability to increase in numbers. Therefore action was taken to place old untreated wooden fence-posts out for the beetles, and monitor beetle numbers. Stephens Island had been more forested in the past, and historically there would have been more fallen logs available as habitat, so this was not an unreasonable avenue to explore. Another possible limiting factor could be the lack of earthworm availability as a food source. This was also to be investigated, together with the feasibility of translocating individuals to set up a new population. The island is free from introduced mammalian predators, so predator control was not a concern. The beetles have been found to shelter under the fence-posts, and the population is being monitored. It is too early to determine whether the population is recovering due to this action, but that beetles have readily taken to using the fence-posts suggests that cover from old logs is a limiting factor for them on the island (I. Millar pers. comm.).

Achievements under the plan

The production of a recovery plan for carabid beetles has primarily ensured that a focussed, rather than ad hoc, approach has been taken to conserving these beetles. It forced people to think about the issues and possible solutions, to consider exactly what is known about each species and to debate and define goals and objectives. These goals and objectives had to align with the wider Departmental strategic goals and objectives. The actions required to achieve those objectives were outlined, and a priority order was assigned to the species covered under the plan. These steps alone resulted in the narrowing down from 56 taxa of concern, to four taxa with immediate needs for which recovery plans were produced. The process forced a focus on what was achievable given current knowledge.

Often it can be difficult to determine if an action is the result of a recovery plan, or whether that work would have been initiated anyway. Recent work has been undertaken on *Holcaspis brevicula* (Brockerhoff et al. 2005), a carabid that does not have a full recovery plan, but is listed in the recovery document as a species that requires survey. A number of surveys have now been completed for this species. It is most likely that the work would have been done at some time, but the document may have added urgency and sound context to the need. In particular, listing this species in the recovery document may have provided a form of justification to biodiversity managers, and made progression of the work a little easier. Many of the issues are well known by those at the "coal-face" of the work, but pragmatically it is often the managers who need to be convinced in order to approve workplans.

Recovery plans thus serve to focus the work for conservation, and catalyse further work. Since the inception of the threatened carabid beetle recovery plan, a number of measures have been undertaken to further the conservation of this group. Work has been initiated on clarifying the taxonomy of some of the species, in particular to determine if a species is valid, or just a clinal variant. However, if the population of concern is likely to be extirpated before its taxonomic distinctiveness can be confirmed, then management should be initiated to conserve the population as a precautionary measure.

An example is the tag-named *Megadromus* "Eastern Sounds". This taxon occurs on Arapawa Island in the Marlborough Sounds, and on the mainland near the Port Underwood Saddle in the Marlborough region. The population on the mainland is very small. There is a noticeable body size difference in beetles of the two populations, and a dilemma arose over whether this tag-named taxon was in fact two distinct species, with taxonomists believing that it was probably just a single taxon. Based on this opinion, management targeted the larger population on Arapawa Island, which was also the more feasible to manage. But if the assumption that it was just a single taxon was wrong, and there were in fact two species present, then there was a risk of losing the mainland "species". Management of the mainland population would have posed significant difficulties. It was very sparse, and practicalities of managing predators would be expensive and labour intensive. There was an obvious need to clarify this situation, to ensure that the actions being taken were adequately protecting the taxon—or taxa. Based on both morphological and genetic evidence the isolated populations represent a single species.

Taxonomy for the sake of taxonomy is not promoted in the plan. If a species is readily identifiable, despite not being formally described, then there is no need to have the taxonomy resolved—it is not a barrier to conservation. The money would be better spent on initiating recovery action for the species.

Since the inception of the recovery plan, progress has been made on all four of the taxa listed as priority species.

Zecillenus tillyardi has had signage erected at its single known site, and improved public awareness in general. It occurs in an area of high recreational use, and habitat damage through public activity was a concern. The site has been closed off to vehicles since the early 1990s (although some illegal off-road vehicle activity still occurs), and there has been a significant change in the vegetative cover. These changes are being monitored to determine if they are beneficial or detrimental to the beetle.

Mecodema laeviceps had not been seen since 1964 and an intensive and extensive search was being planned. However, very similar specimens were found at new sites as part of independent surveys. Whilst there are some differences between these species and the typical *M. laeviceps*, this variation is consistent with normal north–south clinal variation, and it is believed to be insufficient for them to be considered separate species. Low-level efforts are now focussing on determining the beetle's range and what clinal variation exists.

Lessons learnt—issues with current recovery planning

Progression of carabid beetle conservation objectives is still driven by a few dedicated individuals, rather than becoming fully integrated into the Department's work programmes. This is partly because recovery plan actions, whilst having the sign-off and approval of senior management, are not mandatory, and other work can be assigned precedence. There seems little point in going to the effort of producing a recovery plan if the required work is not going to be undertaken. The reality is that not every threatened species can be followed up effectively, so the focus must be on the most urgent cases and areas where an actual difference can be made. The carabid beetle recovery plan attempted to address this as far as possible by prioritising the 56 species taxa that were listed by the DOC as threatened at the time, and narrowing this down to just four taxa. This was not to say that the other 52 taxa do not need some form of conservation work, as undoubtedly all would benefit, but only a few were deemed sufficiently endangered that work must be initiated urgently to ensure their survival in the immediate future.

Having tasks prioritised and assigned to workplans on a regional basis can also create problems, as this means that even if they are deemed important nationally, there is a possibility that regional priorities will take precedence. Currently there is no centralised prioritisation of activities across the Department. Such a system, which could prioritise within as well as across areas within the Department, is a key area to address (e.g., prioritise between the various actions required in the areas of species, pest, recreation, and historic management). A first step is to develop a system that can prioritise actions between species (i.e. is an action for one species a higher priority than an action for another species?), and assure the necessary resources prior to sign-off of a recovery plan. This will require a robust business case to be presented for consideration.

The single species approach to species conservation was a necessary first step as the Department grappled with the threats facing species, the management methods that could be applied, and the complex interactions with other species. Initially this approach worked well. Initially, there were few species with recovery plans, and they were generally high profile politically driven programmes that were always going to be worked on (e.g. iconic birds like the kiwi and kakapo). However the Department now has more recovery plans and actions than can be resourced. The actions covered tend to represent the ideal rather than the practical. As in other parts of the world, there is a need to rationalise these actions, and focus on what *has* to be done to secure the species, rather than what people would *like* to be done.

The Department will also never be able to implement individual recovery plans for all of New Zealand's 2,155 currently listed threatened invertebrate taxa. This list will continue to grow as knowledge of the species and the threats they face increases. The single species approach to species recovery is not a practical stand-alone or long-term approach.

Even the growing number of multi-species recovery plans, which cover groups of taxa sharing a common threat, habitat or management requirement, could not hope to address the concerns of all threatened species within a reasonable timeframe and the resources available.

A more difficult aspect to address is staff motivation to work on the species of concern. Many people have their "pet" species, the species they will go the extra mile for. It may be the reason they decided to work in conservation. So if upon landing their "dream job" they find out that they cannot work on the species they want to, there may be a motivation problem to address. This is difficult to manage once a person is employed.

It may also be that the species they are asked to work on takes them out of their comfort zone or skill set. If they know nothing about it, they may be reluctant to initiate work on it for fear of doing something wrong. However, if clear national priorities are outlined for species recovery, and Conservancy and Area Offices have a clear idea of the work they were expected to undertake, then future recruitment could be targeted to attract those with interests and skills best aligned to the work required.

The way forward?—improvements that could be made

The Department continues to move towards integrated site management, controlling multiple pest and weed species for the benefit of multiple threatened species. There will still continue to be species that will require individual attention due to the factors that affect them, such as unique threats or unique location.

A simple process can be followed addressing the following key questions.

- What species do you want to protect? (a prioritisation process)
- What is causing their decline?
- Can we manage the agents of decline?
- What are the key sites? (a prioritisation process, including identifying areas that will conserve the greatest number of threatened species for the least amount of resources)
- Does it constitute a priority for funding? (a prioritisation process)

Addressing these questions goes a long way to improving some of the drawbacks of the current system.The flowchart below outlines one possible process that could be followed to rationalise species recovery planning within the Department. It should be noted that the process outlined here does not necessarily reflect the beliefs of the Department, but is rather the author's perspective on a possible way forward (Fig. 1).

Descriptors for steps in the flowchart

"Species prioritisation system" and "Species meets criteria for management" steps

The first requirement is to determine which threatened (carabid or other) species are going to actively be managed. This requires further prioritisation of the Department's list of threatened species. Whilst the threat list provides a list of species most at risk of extinction, the sheer number of species in each category means that further prioritisation is required to arrive at a list of species to provide direct management for. Threat to extinction does not always align with those that *can* be managed. Factors likely to be considered along with the threat status include whether the threatening agents are known, and whether a management technique exists to enable some mitigation of these threats. Species that ideally would be managed, but cannot be as there is currently no effective and efficient means, will come under a research loop, described later in this paper.

"Determine management goal" step

The goals of recovery need to be clearly defined. There is always a tension between continuing with the gains made in the recovery of one species, and moving to focus on another species requiring attention. Ideally all species would be brought up to the same level of security. At their broadest definition, these levels of security can be viewed as "securing from extinction", "protecting a representative range", and "maximising genetic diversity". The aim is to have an achievable common goal for all threatened species. Securing from extinction is the obvious first goal to be achieved for all threatened species, before moving on to further goals of protection of a representative range and maximising genetic diversity.

Individual species, management fact sheets created

For each species that has been identified as desirable to manage, a profile should be produced outlining ecological and biological requirements, plus the agents of decline and recommended mitigation methods. These will come into play later when formulating management programmes for a site.

Species distribution mapping

Once species have been prioritised, the next step is to map and overlay their distributions. This will allow the identification of "hot spots" of threatened species—areas or habitats where there are overlaps or alignments of distributions. It will also identify areas where there are no overlaps or alignments, possibly indicating a species that will need to be managed individually. At this stage it is worthwhile overlaying the distributions of threatened species that did not receive priority for management. Whilst they will not be the specific target of the management activities, they may well benefit from those actions, and this aspect should be considered.

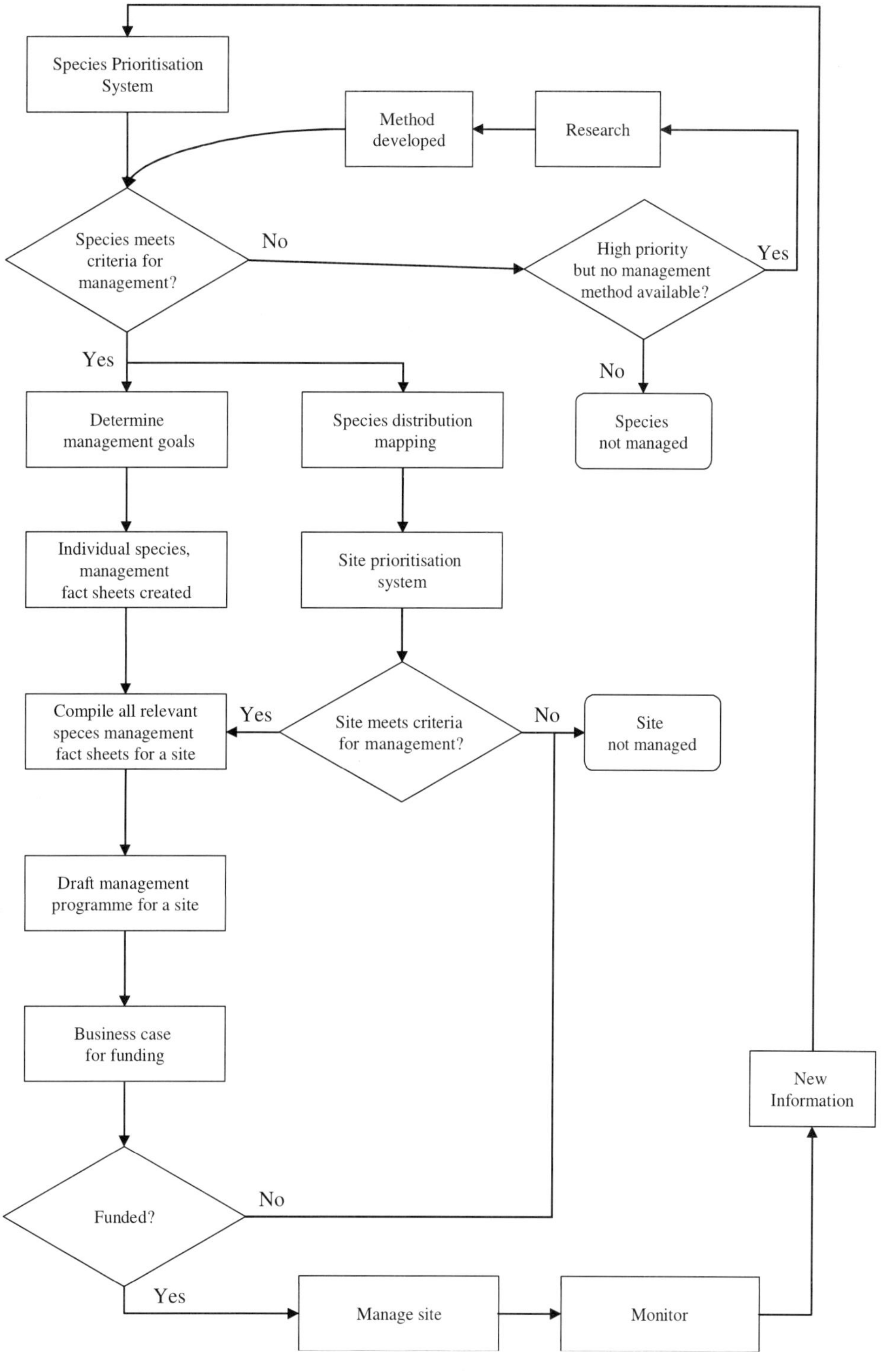

Fig. 1 Decision making process for integrated site management flowchart

"Site prioritisation system" and "Site meets criteria for management" steps

Having done the mapping the next step is identification of a range of sites that will enable the most efficient management, covering a representative range of threatened species. This is a crucial step, and may necessitate trade offs between sites—a slightly less suitable site being chosen because it is more feasible to manage there, or because it contains 10 threatened species as opposed to four. It may turn out that the optimal sites do not occur on land administered by the Department. In this case there will need to be either a collaborative, interagency initiative to manage the site, or consideration of accepting a less optimal site for management on conservation land. Ideally all agencies concerned with species conservation and threat mitigation would collaborate on which sites they place the most value on, and look for synergies and opportunities for gains through collaborative management. Factors such as feasibility, practicality, cost benefit, accessibility, strategic fit, land tenure, public acceptance and funding ability may all be relevant.

Compile all relevant species management fact sheets for a site

Once sites are determined, the individual profile sheets for each species within that site can be looked at with a view to determining what management regime needs to be undertaken in order to address all of the identified threats facing each species. Many of the threatened species present may have similar threats and will benefit from similar management options. However, there may be instances where one management technique is detrimental to another species at the site. The implications of the various management options and any potentially detrimental management practices may need consideration through adoption of a different methodology.

Draft management programme for a site

Having considered all of the threats and recommended control methods, a management programme can be drafted for the site. This should list a few specific actions to achieve the primary goal of preventing extinction. It should also clearly state when management should be scaled-back or stopped.

"Business case for funding" and "Approval of funding" steps

A robust business case should be prepared for submission to senior management for approval to fund the proposed management programme. The business case will follow usual Departmental practice for consideration and approval, as key step in the process. Approval of the business case would also serve as approval of the management programme. Approval would include endorsement of the management actions as well as confirmation of resources.

"Manage site", "Monitor" and "New information" steps

These represent the steps taken once a management programme is approved, through undertaking the actions identified in the management programme, monitoring the success of that programme, reporting and adjusting management as necessary, and adapting management as new information comes to light.

"High priority but no management method available", "Research" and "Method developed" steps

Some of the species that do not meet the management criteria may have been excluded because there is no method currently available by which to manage their agent of decline. If this is the case then there is the option of seeking funding to initiate research to develop effective control tools. Once a method has been developed, the species can be reconsidered for management.

Discussion

The reality for beetle conservation in New Zealand, and invertebrate conservation in general, is that it is not practical to attempt to individually manage every threatened species. A site-based approach is therefore highly desirable. Integrating the management of beetle species with other threatened species also provides a number of synergies, in that many species will benefit through the integrated approach. The integrated approach also considers potential negative effects of management options on other species in a far more structured and objective way than single species management plans do —if they consider this at all.

In a government organisation, accountable to the public, obtaining funding for beetle conservation is always going to be a challenge. In regards to public perception beetles are very low priority in comparison to the so-called "charismatic mega-fauna", in New Zealand, birds. They are often the most visible and audible components of our native ecosystems. In a parallel to "the squeaky wheel gets the oil", the noisy bird gets the funding; the quiet unobtrusive beetle does not, and may be regarded widely simply as a pest. If beetle conservation can find a way of "piggy-backing" on to other conservation initiatives, it will be working more effectively.

Moving to site-based, multi-species recovery, and managing the full suite of threats (animal pests, weeds, and social), ensures that programmes have the greatest chance of providing benefit to a wide range of species, and hopefully maintaining fully functional ecosystems. The costs of such a management regime will of course be greater than that for a single species approach to management. But the gains achieved should justify this increased cost, as the benefits of an integrated approach should outweigh the cumulative benefits of a single species approach (time, monitoring and reporting will validate whether this is correct or not). Essentially, when operating in a resource-constrained environment, there is a need to maximise the conservation gains achieved relative to the investment of resources. Management of all individual beetle or other invertebrate species is not feasible not only because they are so numerous, but also because of information limitations. Many of New Zealand's endemic invertebrates are undescribed, and basic biological or ecological information is not available on them. It is impossible to tailor a management programme to their needs, because their needs are not known. However, when taking a community or ecosystem based approach, a guild level of consideration (e.g. detritivores, saprophytes, herbivores, predators etc) may be useful. The generic requirements for each group can then be determined and a management regime established to maximise the likelihood of their recovery. This could be something as simple as preventing removal of fallen logs, or supplementing areas deficient in natural log-fall with introduced material (providing it has been sourced from a similar community).

Before instigating a site-based recovery plan, a good understanding is needed of the drivers of the system to be managed. If the basics, such as minimising site disturbance by introduced species (in New Zealand, by controlling introduced browsers and predators such as ungulates, rodents and mustelids), are sound, the system should benefit. However, the control must cover a range of species and not just one or two. A precautionary approach would advocate the control of all known pests. Over time, and with sound monitoring in place, some of this control may be able to be scaled back as more information is gained on the impacts of individual pests on a system. This is a lengthy process, taking many years to determine, but is a core need for assessing threats to carabids (and other beetles) in New Zealand.

Management on the scale advocated in this paper is a challenging and difficult undertaking, but its implementation cannot wisely be delayed. It may be the major key to assuring the future of New Zealand's native invertebrates.

References

Anon (2000) The times atlas of the world – Comprehensive ed., 10th edn. Times Books, London

Brockerhoff EG, Berndt LA, Jactel H (2005) Role of exotic pine forests in the conservation of the critically endangered New Zealand ground beetle *Holcaspis brevicula* (Coleoptera: Carabidae). N Z J Ecol 29:37–43

Cooper RA, Millener PR (1993) The New Zealand biota: historical background and new research. Trends Ecol Evol 8(12):429–433

Field LH, Jarman TH (2001) Mating behaviour. In: Field LH (ed) The biology of wetas, king crickets and their allies. CABI Publishing, Wallingford, pp 317–332

Hitchmough R, Bull L (compilers) (in press) New Zealand threat classification system lists 2005. Threatened Species Occasional Publication

Klimaszewski J, Watt JC (1997) Coleoptera: family review and keys to identification. Fauna of New Zealand 37

Larochelle A, Lariviere M-C (2001) Carabidae (Insecta: Coleoptera): catalogue. Fauna of New Zealand 43

McGuinness CA (2001) The conservation requirements of New Zealand's nationally threatened invertebrates. Threatened Species Occasional Publication No. 20. Department of Conservation, Wellington

McGuinness CA (2002) Threatened carabid beetle recovery plan (2002–2007). Department of Conservation, Wellington (Unpublished)

Molloy J, Davis A (1994) In: Tisdall A (comp) Setting priorities for the conservation of New Zealand's threatened plants and animals, 2nd edn. Department of Conservation, Wellington

Parrish R, Stringer I, Booth A, Brook F, Sherley GH (2005) Northern giant landsnail recovery plan: *Placostylus* spp, *Paryphanta* spp and *Rhytidarex* sp 2005–2015. Department of Conservation, Wellington (Unpublished)

Sherley GH (1998) Threatened weta recovery plan. Threatened Species Recovery Plan 25. Department of Conservation, Wellington

Walker KJ (2003a) Recovery plans for *Powelliphanta* land snails 2003-2013. Threatened Species Recovery Plan 49. Department of Conservation, Wellington

Walker M (2003b) Short-horned grasshopper recovery plan 2003–2008. Department of Conservation, Wellington (Unpublished)

J Insect Conserv (2007) 11:43–46
DOI 10.1007/s10841-006-9017-x

BEETLE CONSERVATION

Notes on the habitat and adult behaviour of three red-listed *Colophon* spp. (Coleoptera: Lucanidae) of the Cape Floristic Region, South Africa

H. Geertsema · C. R. Owen

Received: 2 October 2005 / Accepted: 15 March 2006 / Published online: 2 November 2006

Abstract The habitat restrictions of three species of *Colophon* (*C. stokoei, C. neli, C. westwoodi*) in the Cape Floristic Province are discussed, together with their patterns of seasonal activity and biological features in captivity, to augment the biological background needed for their effective conservation management.

Keywords Biomes · Conservation · Montane beetles · Protection · South Africa

Introduction

The specialized montane stag beetles in the genus *Colophon* are Red-listed and CITES-listed (IUCN 2004) endemic to the Western Cape Province and extreme western areas of the Eastern Cape Province mountains of South Africa. They are periodically threatened by commercial collecting and possibly vulnerable to global warming.

In a Provincial Proclamation (No. 24/1992) all species of *Colophon* were placed in Schedule 1 of the Nature and Environmental Conservation Ordinance, 1974 of the Province of the Cape of Good Hope (now Western Cape Province). In November 1994, the Netherlands submitted a proposal to include the genus *Colophon* in CITES Appendix I, but it was eventually listed in Appendix III (Anonymous 1994). The intention of this protective measure was ostensibly to protect these insects from commercial exploitation, but in fact only resulted in drawing attention to these beetles and in the creation of a substantial black market (Gess and Gess 1993), adding to the collecting pressure on present populations. Since the early 1990s, another development has led to concern for the survival of Cape stag beetles, namely the construction of repeater stations for mobile telephones, many of the targeted sites coinciding with the habitat of *Colophon* species.

Endrödy-Younga (1988) considered the species in this genus as representatives of a 'mountain relict biome', presently restricted to the higher elevations (1000–2000 m) of the western and southern Cape mountains. This elevated montane biome is regarded as a refugium in to which the beetles have retreated after the low-lying plainland became uninhabitable due to aridification and temperature increase (Endrödy-Younga 1988).

Very little is known of the biology of the 17 *Colophon* species (Bartolozzi 2005). Barnard (1929) reported some general observations on *Colophon*. Beetles live on the summits of the mountain ranges and seem to be mainly nocturnal, or wandering around in misty weather, mostly in the early hours of the morning (6–8 am), hiding by day under stones or amongst low vegetation (Barnard 1929; Bartollozzi and Werner 2004). They are probably root-feeders, feeding on the low bushy Restionaceae always present on the Cape mountains (Barnard 1929; Brinck 1956). More detailed observations on the biology of *C. izardi* Barnard and *C. montisatris* Endrödy-Younga from the Swartberg

H. Geertsema (✉)
Department of Conservation Ecology and Entomology, University of Stellenbosch, Private Bag X1, Matieland 7602, South Africa
e-mail: hge@sun.ac.za

C. R. Owen
212 Helderberg College Road, Somerset West 7130, South Africa

and Langeberg ranges, respectively, were later reported by Endrödy-Younga (1988). In October 1977, Endrödy-Younga found live specimens of *C. izardi* but baited ground-traps failed to attract these beetles. Subsequently, he did obtain live specimens of *C. izardi* and *C. montisatris* which were observed for up to 84 days in the laboratory. The habitat of *C. montisatris* in the Swartberg was limited to a discrete strip of about 400 m horizontally and 20 m vertically, while the *C. izardi* population in the Langeberg occurred at sites 8 km apart and differing by about 400 m in elevation. The latter species is known from five additional localities at increasing distances from the observation area (Endrödy-Younga 1988: Fig 10). The adult activity period for *C. izardi* was from October to February (Barnard 1929; Endrödy-Younga 1988). Endrödy-Younga (1988) commented on the slow undirected movement of males of both species in the field. Aspects of the ecology of *Colophon* species were reported (in Japanese) by Mizukami and Kawai (1997).

No ecological field studies have been done on any of the Cape stag beetles. As a result, no published information is available on population sizes, rates of population turnover, predation rates and ecological requirements, although Bartolozzi and Werner (2004) do refer to their respective conservation status, most regarded as rare. It is thus difficult to assess the conservation status and the threat to these species (Anonymous 1994). All species of *Colophon* have highly restricted ranges, some more than others. Holm in Anonymous (1994) refers to *C. montisatris* as occurring at only one locality of a few ha in extent, and estimates its total population at less than 1,000 individuals.

In view of limited information on the preferred habitat and behaviour of these beetles (Anonymous 1994) and concern over possible requests for erecting repeater stations, the Western Cape Department of Environmental and Cultural Affairs, Directorate Nature Conservation and Museums, granted special permission to collect and study these beetles from specific localities so as to contribute to the existing information pool.

During January 1996, 26 individuals of *C. stokoei* Barnard, 6 *C. westwoodi* Gray and 12 *C. neli* Barnard were collected in the Hottentots Holland Mountains, on Table Mountain and in the Riversdale Mountains of the Western Cape Province, respectively. In all instances, a small number of the beetles were observed and some were collected (to protect populations in the wild, precise locations have been withheld). Live specimens were taken to the laboratory in Stellenbosch, Western Cape, to study their behaviour. Collected specimens are deposited in the insect collection of the Department of Conservation Ecology and Entomology, University of Stellenbosch (USEC).

Habitats of C. stokoei, C. neli and C. westwoodi

Localities for these beetles are usually given as Hottenhots Holland Mountains or Swartberg Mountains (Endrödy-Younga 1988). However, the habitat of these beetles is far more restricted and usually confined to glades at, or near the top of these mountains, but sometimes at considerably lower elevations. The habitat for *C. stokoei* in the Hottentots Holland Mountains (1,150–1,550 m asl), and *C. neli* in the Swartberg Mountains (<1500 m asl), consisted of sloping flatlands, ranging from horizontal to about a 30° inclination. These were covered with typical, but rather short, montane fynbos vegetation, interspaced with rocks. The habitat (1,000 m asl) of *C. westwoodi* was fairly marshy flatland, with taller restioid vegetation (30–50 cm; dominated by *Elegia* spp.) on dark, poorly-drained soil presumably with a high humus content. In 1992, specimens of *C. westwoodi* were also observed near the De Villiers Reservoir on Table Mountain in a similarly vegetated habitat. During one visit in mid January 1996, 12 trampled specimens of *C. westwoodi* were found on a path often used by tourists, close to a favourable habitat. On Table Mountain, little evidence of lizard predation was found, perhaps due to the unsuitability of the habitat for lizards. However, a scat of a predator, most probably that of the Small Gray Mongoose [*Galerella pulverulenta* (Wagner)], containing the remains of *C. westwoodi*, was found.

The habitat of *C. montisatris*, at 2,085 m asl, consisted of a tiny area of about 400 m horizontally and 20 m vertically in the Swartberg mountain range. Outside this area not even fragments of this species could be found and it is likely that this species is on the verge of natural extinction unless the regional climate soon turns more humid and cooler. In the Langeberg Mountain range (at between 1300 and 1600 m asl), *C. izardi* was found at sites 8 km apart and differing in altitude by about 400 m. Of the two ranges, the northern (Swartberg) is the drier, with less frequent cloud cover, higher summer temperatures and a long dry spell in summer. Environmental differences, including that of vegetation, soils, exposure, etc., are considerable between these two habitats (Endrödy-Younga 1988). Although *Colophon* species are confined to certain localities, *C. oweni* occurs sympatrically with *C. izardi* on the east side of Tradouw's Pass in the Riversdale Mountains of the Western Cape Province and *C. thunbergi* is found on the west side of this pass (Bartolozzi 1995). The contribution by Mizukami and Kawai (1997) provide good pictures of the most of the habitats of various *Colophon* species; unfortunately, their detailed ecological notes on the genus remain obscure, being presented in Japanese.

Seasonality

The adult activity period for *Colophon* ranges mainly from October to March (Barnard 1929; Endrödy-Younga 1988, own observations), i.e., the whole of summer. Females were rarely collected: five males and one female *C. westwoodi*, 12 males *C. neli* and 3 females and 23 males *C. stokoei* were collected. The ratio of females to males is extremely low. Possibly males are more active and thus more easily spotted than females, or sex ratios fluctuate during the season. It is also possible that females spend more time in hiding, while males search actively for females. In view of the long emergence period and life span, this is the most likely explanation. All beetles were collected in the middle of their emergence period, and it may well be that females appear earlier than males. Endrödy-Younga (1988) found specimens, especially females, with strongly eroded fore tibiae, indicating a long life span similar to that shown in captive Tenebrionidae.

A *C. stokoei* larva was found about 20 cm depth in a mixture of washed-down soil and nearly completely rotten plant material, retained by and lodging against the rocks on Table Mountain. Rearing it to the adult stage proved unsuccessful. Two third-instar larvae of *C. neli* were previously found by one of the authors (CRO) in humus-rich soil. The larva of *C. neli* was described subsequently by Scholtz and Endrödy-Younga (1994).

Activity patterns

Adult beetles generally became active from late afternoon (17 h) to early nightfall (20 h), but appeared earlier when mountains were covered in cloud, or misted over due to the south-easterly winds experienced during the hot, dry summer months in the Western Cape Province. When inactive, beetles hid in the vegetation adjoining vertical rock faces and were sometimes found under rocks. Movement in vegetation was slow, making the observation of beetles difficult. However on open ground movement was fairly rapid and occurred in random directions until shelter was found. This concurs with the observations of Endrödy-Younga (1988).

Colophon species studied elsewhere have also been found to be decidedly diurnal in activity. Specimens of *C. izardi* have been observed in thick fog after sunset or at about 16 h in drizzling rain and biting cold wind (Endrödy-Younga 1988). Most specimens have been seen during the late afternoon hours, often after sunset but before dark (Endrödy-Younga 1988). This contrasts with the observations of Barnard (1929) who collected most of his specimens between 6 and 8 h and at 14 h in the hot midday sun.

Behaviour and longevity of captive beetles

Beetles collected in the field were placed outdoors in 5l plastic containers, half-filled with compost-rich garden soil during mid January 1996. Shelter, provided by supported tile shards, was rapidly occupied by beetles during the day. Locomotory activity, at ambient temperatures of 17–22°C, commenced from 17 h onwards and was directed away from the provided shelter. Males encountering other males would attempt to push them aside, or use their mandibles to attack them. Many specimens, both in the field and in captivity, had lost parts of their legs and antennae during such encounters. When confronted by a female, mating was attempted. The male, or males, would approach the female from the rear, using their strong fore legs as pincers to hold the female between the pro- and mesoterga. Coupling lasted from a few minutes to more than two hours. However, no eggs were produced. Endrödy-Younga (1988) reported that attempts were made to provide food, including a range of plants from their habitat, but none of the captive specimens were observed to feed, unless on the vegetation debris that was richly supplied in the soils brought with them (Endrödy-Younga 1988). Whilst conducting our research, beetles were only interested in droplets of water. No feeding was observed on the supplied compost from their localities, but one beetle showed some interest in a slice of apple. It could well be that in nature they ingest organic soil matter as it is most unlikely that these beetles would survive for months without food.

Colophon westwoodi beetles, collected on 6 January 1996, died after 18–20 days (2 females) and between 23 and 32 days (4 males)(mean 26.25 days), respectively. Six males of *C. neli*, collected on 7 January 1996, died after 24–32 days (mean 26.50 days), respectively. *C. stokoei* were collected on 1 and 22 January 1996, of which males died after 11–56 days (mean 32.48 days, $n = 23$) and females after 29–61 days (mean 41.67 days, $n = 3$). Endrödy-Younga (1988) kept *C. montisatris* alive for 84 days.

Conservation efforts to protect these flagship beetles from collectors have so far relied on legislative means and the fact that they occur in remote montane regions. Further biological data on *Colophon* are now needed for a improved documented account of population sizes, geographical ranges, habitat requirements and population dynamics.

Acknowledgements We thank JH Neethling, former Chief Director, Directorate Nature Conservation and Museums, Department of Environmental and Cultural Affairs, Western Cape Province, for a special permit (No. 266/95) to collect and study *Colophon* beetles; P Gildenhuys (Chief Law Enforcement Officer, Cape Nature) for literature provided. We also thank all reviewers of the manuscript for their input.

References

Anonymous, 1994. Analyses of proposals to amend the CITES Appendices. Ninth Meeting of the Conference of the Parties, Fort Lauderdale (USA), 7–18 November 1994. Prepared by the IUCN Species Survival Commission TRAFFIC Network

Barnard KH (1929) A study of the genus *Colophon* Gray. Trans R Soc S Afr 18:163–182

Bartolozzi L (1995) Description of a new species of *Colophon* from South Africa (Coleoptera, Lucanidae). Fragm entomol Roma 26(2):333–340

Bartolozzi L, Werner K (2004) Illustrated catalogue of the Lucanidae from Africa and Madagascar. Taita Publishers, Hradec Kralove, Czech Republic. 189 pp

Bartolozzi L (2005) Description of two new stag beetle species from South Africa (Coleoptera: Lucanidae). Afr Ent 13(2):347–352

Brinck P (1956) Coleoptera: Lucanidae. S Afr Animal Life 3:304–335

Endrödy-Younga S (1988) Evidence for the low-altitude origin of the Cape Mountain Biome derived from the systematic revision of the genus *Colophon* Gray (Coleoptera: Lucanidae). Ann S Afr Mus 96:359–424

Gess FW, Gess SK (1993) Irresponsible collecting for financial gain. Lett African Wildlife 47:187

IUCN, 2004. 2004 IUCN Red List of Threatened Species http://www.redlist.org

Mizukami T, Kawai S (1997). Nature of South Africa and ecological notes on the genus *Colophon* Gray. Gekkan-Mushi Supplement 2:1–79. (in Japanese)

Scholtz CH, Endrödy-Younga S (1994) Systematic position of *Colophon* Gray (Coleoptera: Lucanidae), based on larval characters. Afr Ent 2(1):13–20

J Insect Conserv (2007) 11:47–59
DOI 10.1007/s10841-006-9018-9

BEETLE CONSERVATION

Are we doomed to repeat history? A model of the past using tiger beetles (Coleoptera: Cicindelidae) and conservation biology to anticipate the future

David L. Pearson · Fabio Cassola

Received: 23 September 2005 / Accepted: 20 December 2005 / Published online: 7 November 2006

Abstract Studies of conservation biology involving tiger beetles have become increasingly common in the last 15 years. Governments and NGOs in several countries have considered tiger beetles in making policy decisions of national conservation efforts and have found tiger beetles useful organisms for arguing broad conservation issues. We trace the evolution of the relationship between tiger beetle studies and conservation biology and propose that this history may in itself provide a model for anticipating developments and improvements in the ability of conservation biology to find effective goals, gather appropriate data, and better communicate generalizations to non-scientific decision makers, the public, and other scientists. According to the General Continuum of Scientific Perspectives on Nature model, earliest biological studies begin with natural history and concentrate on observations in the field and specimen collecting, followed by observing and measuring in the field, manipulations in the field, observations and manipulations in the laboratory, and finally enter theoretical science including systems analysis and mathematical models. Using a balance of historical and analytical approaches, we tested the model using scientific studies of tiger beetles (Coleoptera: Cicindelidae) and the field of conservation biology. Conservation biology and tiger beetle studies follow the historical model, but the results for conservation biology also suggest a more complex model of simultaneous parallel developments. We use these results to anticipate ways to better meet goals in conservation biology, such as actively involving amateurs, avoiding exclusion of the public, and improving language and style in scientific communication.

Keywords Cicindelidae · Conservation biology · History · Models · Tiger beetles

CXLV, Studies of Tiger Beetles

D. L. Pearson (✉)
School of Life Sciences, Arizona State University, Tempe, AZ 85287-4501, USA
e-mail: dpearson@asu.edu

F. Cassola
Via F. Tomassucci 12/20, I-00144 Rome, Italy

Introduction

The early 20th Century Spanish philosopher, George Santayana, is credited with the quotation, "Those who cannot learn from history are doomed to repeat it." Although easily dismissed as a trivial aphorism, is it possible that this statement constitutes a testable hypothesis that we can use to understand and anticipate advances in sciences such as conservation biology?

Conservation biology is a field with too few years of experience to have engendered broad interest in its past (Zirnstein 1996; Meine 1999; Siemann 2003). However, its history together with that for longer-established supporting fields, such as systematics, genetics, wildlife management, and ecology, may hold critical information for developing future directions and goals for conservation biology. Faced with constant shortages of funding to adequately gather information and conduct studies, lessons from history may be useful as another set of tools in the quest for meeting these goals (Maienschein 2000; Gaddis 2004).

In a search for patterns within the history of scientific studies, historians have analyzed several fields from physics (Nye 1996) to biology (Killingsworth and Palmer 1992). Are there steps common to all scientific endeavor? What recognizable patterns of change take place, and what are the significant factors causing the changes? How can they best be compared? Apart from satisfying intellectual curiosity, a solid understanding of patterns in the development of science could prove useful for conservation biology in many ways. It could: (1) help determine priorities for funding agencies, (2) enable biologists to better communicate with and inform non-scientific decision makers, (3) focus individual researcher goals, (4) prepare cooperative research agendas, (5) formulate more reliable and efficient models for management and conservation goals, and (6) help anticipate problems that can then be ameliorated.

Methods

The historical model

History does not lend itself to experimental repeatability (Gould 1989), and thus tests of patterns in history rely on alternative methods. One of the most reliable techniques for answering pertinent historical questions and testing for patterns is by using insights from one field to tell us something about another—a process called consilience by historians. In so doing, we can make sense of the past and perhaps anticipate the future (Gaddis 2004). Within biology, such patterns have been proposed for understanding the historical progression of human cultures. Important causes with consistent outcomes across unrelated cultures include environmental factors (Rolett and Diamond 2004), plant and animal domestication (Diamond 2002), disease (Acemoglu et al. 2001) and food production (Hibbs and Olsson 2004).

Along the same lines, one general model of the history of science proposed to anticipate historical patterns in biology is the General Continuum of Scientific Perspectives on Nature (GCSPN) (Killingsworth and Palmer 1992). According to the GCSPN, earliest biological studies begin with natural history and concentrate on observations in the field and specimen collecting, followed by observing and measuring in the field, manipulations in the field, observations and manipulations in the laboratory, and finally enter theoretical science, including systems analysis and mathematical models. What is not clear is whether each of these chapters in the development of science can be identified as a chronological step or phase, or even more controversial, whether each step has identifiable and quantifiable characters that can be used to estimate the maturity of the field of study (Farber 2000).

In addition to these uncertainties, this model has other constraints. As with all models, simplification is an acceptable aspect of their use as long as the results are interpreted within these limitations. Also, similar to many ecological and landscape studies, using time intervals that are too small or too large can obscure important patterns. Finally, sociological, economic, and psychological forces can be more crucial in affecting models of temporal changes than generally realized, but these factors often are difficult to incorporate into general models. With these assumptions in mind, Battalio (1998) listed a series of specific characters that would demonstrate historical steps within the GCSPN model:

(STEP 1) Descriptive natural history and search for new species predominate
(STEP 2) Now an experimental science rather than a natural history model
(STEP 3) Power is transferred from expert amateurs to trained professional scientists, and graduate training for employment in the field has become available
(STEP 4) Systematics no longer dominant, and research focused more on theoretically complex issues with extensive use of graphs and statistical inference in publications
(STEP 5) Formation of research teams and increasing evidence of socialization, such as use of acknowledgments sections, associations of peers, and co-authored publications
(STEP 6) Technical terminology and methodology so refined they now limit the audience that can fully comprehend them (Fig. 1).

Test subjects

The history of entomology provides a rich and varied set of potential subjects to test the model. However,

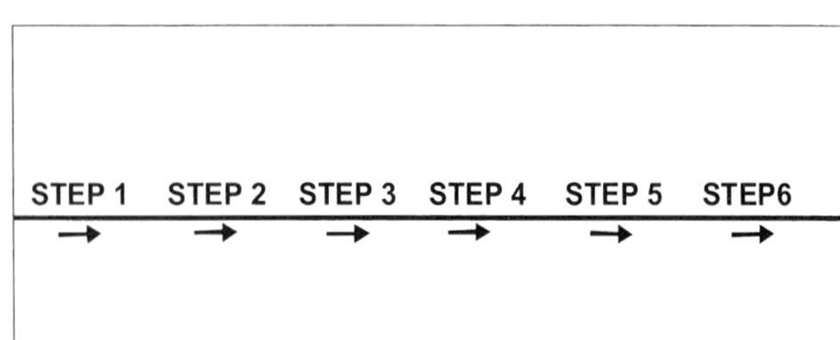

Fig. 1 Linear progression of steps in GCSPN model in which each step replaces the former one

because of the myriad and often independent histories of various insects groups within entomology (Sorensen 1995), we felt that an initial test of the GCSPN model would be more manageable by using a single group. Tiger beetles (Coleoptera: Cicindelidae) provide a relatively discrete taxonomic unit whose history is well documented (Pearson and Cassola 2005). The tiger beetles are a small but distinct group of over 2600 species whose biology is also well known (Pearson and Vogler 2001). These beetles are attractive, fast-flying and fast-running insect predators that occur in many diverse habitats around the world. Many of the same characteristics of tiger beetles that have generated considerable interest among amateurs and professional biologists have also contributed to their increasing role in conservation studies. Most important among these characteristics is the ease with which most species can be found and identified in the field, their habitat specificity, and their value as indicators of habitat health and of biodiversity. Also, because they have been well-collected and studied, their past and present distributions are known sufficiently to evaluate historic trends of decline in range or abundance (Desender et al. 1994; Knisley and Fenster 2005).

In addition, we will use the history of the field of conservation biology as a test subject. With a combination of narrative and comparative analysis, we propose to compare these two histories to test the validity of the model. Finally, we pursue the possibility that if the resultant pattern of steps conforms to the GCSPN model, can the model and its assumptions be used to anticipate and direct future steps in conservation biology?

Results

Step 1: Descriptive natural history and search for new species predominate

As claimed by the GCSPN, much of the earliest history of conservation biology revolved around documentation of species, in this case their extinctions. In the late 18th Century, American authors Ralph Waldo Emerson and Henry David Thoreau influenced the development of Transcendentalism, a philosophy associated with nature. Through their writings, preservation of nature and wilderness became a powerful, novel doctrine. In the midst of manifest destiny and impressions of inexhaustible resources, the unexpected disappearance of once abundant species, such as the Passenger Pigeon, and near extinction of the American Bison, first made extinction seem a real possibility, and the causes of extinction of individual species became an important area of study for the nascent field of conservation biology.

Because of an extensive knowledge of taxonomy and distribution starting with Linné (1758), tiger beetles lent themselves to early studies of declining populations and extinctions. As such, several species and populations of tiger beetles became some of the first insects declared legally endangered or threatened with extinction.

Pearson et al. (2005) estimate that at least 33 (15%) of the 223 named species and subspecies of tiger beetles in Canada and the United States may be declining at a rate that justifies their consideration for inclusion on the US Fish and Wildlife Service's List of Endangered and Threatened species (Fig. 2). However, at present, only four of these are officially listed by the federal government, and several others are under consideration for listing. In addition, several other countries (Belgium, Canada, Germany, Great Britain, Lithuania, The Netherlands, South Africa and Sweden), at least 24 individual states and provinces within the United States and Canada, and international NGOs (World Conservation Monitoring Centre and IUCN) have developed lists of endangered and threatened species that include tiger beetles.

Few insects are well-enough known globally to document these types of population decline. Because of the rich collections of tiger beetle specimens available for study, however, the disappearance of species from former parts of the range can be authenticated. From these historical records, some long-term changes in the environment can also be deduced (Nagano 1980; Desender and Turin 1989; Desender et al. 1994; Yarbrough and Knisley 1994; Kamoun 1996; Trautner 1996; Berglind et al. 1997; Diogo et al. 1999; Knisley and Hill 2001; Richoux 2001; Sikes 2002; Goldstein and Desalle 2003; Horgan and Chávez 2004; Mawdsley 2005). Thus, tiger beetles help offer a window into our past and can provide insight as to where protective measures are needed (Babione 2003).

Step 2: Now an experimental science rather than a natural history model

For tiger beetle studies, the major intellectual advance during the last half of the 18th Century was an often-conflicting attempt to place the growing number of species into a natural array of groupings. By moving from pure description to evolutionary questions, these attempts at phylogenetics were also some of the first signs of a change into an experimental paradigm (Barrow 1998). With more species known, better

Fig. 2 Controlled area in Santa Cruz Co., California, to protect the officially endangered Ohlone Tiger Beetle (*Cicindela ohlone*) Photo courtesy Univ. Calif. Santa Cruz Grounds Dept.

chances for comparisons, and greater competition for research subjects among the increasing number of experts, tiger beetle systematists ventured into more sophisticated areas of research. Field naturalists such as A.R. Wallace and H. W. Bates often collected tiger beetles wherever they traveled. Emergent but significant ideas about behavior, ecology and evolution also grew from their experiences of collecting and observing these beetles. The German medical doctor, Walther Horn, became the greatest authority and acknowledged specialist of the tiger beetle family, working almost solitarily for more than 50 years. Although predominantly taxonomic in nature, his articles began, later in the 1900s, to incorporate experimentally testable ideas of habitat, biogeography and intraspecific variation (subspecies).

Besides reconstructing the past, tiger beetles are useful for conservation in other ways. Because of political, sociological and economic pressures, conservation policy and research are under pressure to produce quick results. This pressure is so pervasive, and the time, money and personnel to do the work are so limited that conservation biology is called a "crisis discipline," in which risk analysis has become a major element (Maguire 1991). A common approach to resolving these problems has been to use indicator taxa as test organisms that purportedly represent other taxa in a complex environment. By focusing studies on a small but representative subset of the habitat or ecosystem, patterns of habitat degradation and population losses can be more quickly and clearly distinguished (Noss 1990).

Unfortunately most taxa suggested for use as indicators have been selected primarily on the basis of their public appeal (Pearson 1994). The consequences have cast doubt on the general usefulness and accuracy of bioindicators in conservation policy-making. For instance, among animal taxa, most studies using indicator taxa have relied on vertebrates, especially those "species of high public interest" (USDI 1980). Vertebrates, however, tend to be relatively long-lived, have low rates of population increase, long generation times, and comparatively low habitat specificity (Murphy et al. 1990), all of which tax the time and finances for proper investigation. As a result, there is a trend now to rely more and more on arthropod species, especially insects, instead of, or in addition to, vertebrates as appropriate indicator taxa (Pyle et al. 1981; Kremen 1992; Samways 1994; McGeoch 1998).

Tiger beetles have been used throughout the world to test and develop better guidelines for choosing bioindicators (Holeski and Graves 1978; Schultz 1988; Bauer 1991; Pearson and Cassola 1992; Rivers-Moore and Samways 1996; Kitching 1996; Rodríguez et al. 1998; Cassola and Pearson 2000; Cassola 2002; Arndt et al. 2005). First, the category of bioindicator is determined (Kremen et al. 1993). Will it be used for monitoring (Greenberg and McGrane 1996), in inventory (Lees et al. 1999), as an umbrella (Mittermeier et al. 2004), or some other type of model organism? Then a claim is made that a species or taxon, such as tiger beetles, is ideal as a bioindicator in a specific category. That leads to tests of whether this proposed indicator taxon meets the demands of widely accepted logistical and biological criteria for ideal indicators within each category. A useful bioindicator taxon should have characteristics such as stable taxonomy, well-known biology and readily observed and manipulated (Brown 1991). More recently, it has become evident that even when chosen carefully, a single taxon is unlikely to be adequate. Seldom will a single taxon reflect accurately an entire habitat or ecosystem (Ricketts et al. 1999). Choosing a suite of indicator taxa from different trophic levels or different subhabitats within the area of interest probably produces better data on which to base rational and informed biological and policy decisions. Nevertheless, each of the suite of candidates should be vetted experimentally to determine its appropriateness for that specific use as a bioindicator.

Step 3: Power is transferred from expert amateurs to trained professional scientists, and graduate training for employment in the field has become available

In the late 1800s, the first conservation organizations, such as the Audubon Society and the Royal Society for the Protection of Birds, were formed with both professional and amateur participants. In the next few decades, the work of these professionals and amateurs created many conflicts, such as the benefits of specimen collecting and use of common names. Little by little, professional academicians and government employees with advanced degrees, such as Aldo Leopold and Rachel Carson, took over the study and communication of conservation problems. In 1985 the Society for Conservation Biology was established, and by 2000 it had 5100 professional members. Conservation studies that involve insects have become more common (Bossart and Carlton 2002) in recent years, often focused by international insect organizations, such as the Xerces Society for the Conservation of Invertebrates. In 1997 the Journal of Insect Conservation was launched in conjunction with the British Butterfly Conservation Society. By this time additional national societies dedicated to the conservation of insects had been formed in Asia, Europe and North America. Along with journals focused on this area, graduate programs and salaried positions as conservation biologists, many of whom use insects as test organisms, became established, and the leadership and predominance of professionals became more and more obvious.

For tiger beetles, the near monopoly of a single expert, Walther Horn, had great influence on the direction of studies. Beyond his tight control of tiger beetle taxonomy, however, a few other professional biologists began to publish scientific articles using tiger beetles as test organisms for geological history (Wickham 1904), ecology (Shelford 1907), and behavior (Shelford 1902). The use of tiger beetles in conservation did not begin until the late 20th Century (Pearson and Vogler 2001), and some potentially divisive problems, such as the development of common names, were less disruptive among amateur and professional tiger beetle workers (Pearson 2004) than with other groups, such as birds, butterflies and dragonflies.

Even more subtly, professionalization of scientific articles, including those for conservation biology and tiger beetle studies, is reflected in its evolving language, writing styles, and grammar. Linguistic analysis of journals and scientific articles shows consistent changes that indicate levels of expertise and establish levels of authority, further separating professionals from amateurs. Some examples of changing words include adverbs that show degrees of reliability, such as "undoubtedly" and "possibly," induction, such as "must" and "evidently," identification of hearsay evidence, such as "it seems" and "apparently," reservations of deduction, such as "presumably" and "could," and hedges, such as "approximately" (Chafe 1986). In addition, professional science writers use distinctive writing devices that include reduced use of personal pronouns, reliance on passive voice, a decrease in the number of simple sentences, the presence of technical terminology, an emphasis on reliability of evidence, and the use of citations (Lakoff and Johnson 1980). Carter (1990) also showed that although professionals rewriting scientific articles for semi-popular or popular consumption tend to write in broader generalities and use methods more similar to amateurs, they retain a concept of domain-specific knowledge that distinguishes them from the style of amateurs.

Step 4: Systematics no longer dominant, and research focused more on theoretically complex issues with extensive use of graphs and statistical inference in publications

Among tiger beetles, in areas other than taxonomy, the 1960s saw a relatively small increase in articles published on behavior, ecology, morphology, biogeography and ecology (Pearson 1988). But starting in the 1980s, physiological studies of tiger beetles emerged (Dreisig 1980; Hadley et al. 1988; Gilbert 1997; Hoback et al. 2000; Okamura and Toh 2004). In the 1990s, genetics studies began to appear (Galián et al. 1990; Proença et al. 2002), and by this time, these and other non-taxonomic publications constituted 85% of the articles on tiger beetles with statistical procedures and graphs.

One area in which tiger beetles were at the forefront of more complex conservation biology studies was in the statistical application of assumptions of dependence among data points. In initial comparisons of species patterns across regions and countries, Pearson and Cassola (1992) claimed that among the tested attributes of tiger beetles as an ideal bioindicator was a high correlation between their species numbers and those of other groups. If one goal is to establish conservation areas with the highest species diversity, tiger beetles were very useful because where you found more of them you also found more of other species like birds and butterflies. But tiger beetles, at the right season, could often be surveyed in a few weeks whereas birds took years to survey adequately in the same area. In addition, it was easy to train students and local workers to observe and sample tiger beetles, but training these same people to observe other taxa, such as birds and butterflies, was an enormous undertaking. Thus, one could argue that tiger beetles are logistically useful and biologically appropriate candidates to help represent entire habitats or ecosystems for species inventories.

A major problem, however, was the misapplication of a common simplifying component in statistical tests used by many biologists (Carroll and Pearson 1998a). In virtually all traditional statistical tests, a datum from one point in space or time is assumed to not influence or affect any another datum in the analysis obtained from a different point in space or time (independent). If, however, the data are dependent (often called autocorrelated), and many subsequent studies show that many if not most biological data are likely to be dependent, the resultant analysis may be faulty or misleading (Carroll and Pearson 2000). Many researchers now apply more appropriate statistics, such as geostatistics (Cressie 1991), in conservation biology that avoid the assumption of independence. Tiger beetles were among the first taxa using these modern analytical techniques (Carroll 1998; Pearson and Carroll 1998; 1999; Carroll and Pearson 1998b, Pearson and Carroll 2001).

In addition to pioneering statistical analyses, tiger beetles also were used in early applications of molecular analysis for geographical implications of conservation. For instance, the subdivision of lineages of the tiger beetle species, *Cicindela dorsalis*, in Florida between the Gulf of Mexico and the Atlantic Ocean, can be detected only with molecular markers. However, the fact that species of several taxa on one side of a barrier are consistently different from those on another is highly significant for conservation (Pearson and Vogler 2001). These regions of distinctive genetic overlap can reflect historical events in evolutionary time (Crandall et al. 2000; Goldstein et al. 2000; Satoh et al. 2004). By incorporating an evolutionary time scale, we not only gain another valuable factor to include in our conservation planning, but it also makes us aware that areas chosen for protection require management goals focused not just on 10, 20 or even 100 years, but for much longer into the past as well as the future (Schwartz 1999; Barraclough and Vogler 2002).

Step 5: Formation of research teams and increasing evidence of socialization, such as use of acknowledgments sections, associations of peers, and co-authored publications

Conservation biology has quickly moved from single and often isolated researchers such as Leopold and Carson to a predominance of interactive researcher teams. Among tiger beetle studies, there is considerable evidence for similar changes on a broad level, some of them apparently caused by the appearance of field guides and general books on the biology of tiger beetles beginning in the 1990s (Knisley and Schultz 1997; Leonard and Bell 1999; Acorn 2001; Choate 2003; Pearson et al. 2005; Pearson and Shetterly 2006). Before this time, only individuals with time and interest to search through often obscure journals and arcane terms could acquire the basic knowledge and identification skills to do research using tiger beetles. More specific evidence of socialization is in co-authored publications. In one of the first general reviews of tiger beetle biology (Pearson 1988), only 23% of the cited articles were co-authored. Twelve years later in a book on general tiger beetle biology (Pearson and Vogler 2001) 40% of its citations were co-authored. In 1969, an informal correspondence among tiger beetle enthusiasts developed into a journal called "Cicindela." Another indicator of socialization showed advances within this highly specialized journal. In the 1970s only 2% of its articles had acknowledgments sections; in the 1980s, 26% had these sections; and in the 1990s, 83% of them did.

Similarly, the complex nature of modern conservation biology research necessitates more and more research teams. For instance, many modern conservation biologists working on rare and endangered species now rely heavily on molecular markers (Avise 1994; Galián and Vogler 2003) to distinguish species and populations within species. The importance of conserving intra-specific variation is reflected in the U.S. Endangered Species Act, which calls for the conservation of "independent population segments". This makes conservation of distinct populations within a species a legal requirement, and involves coordination of field biologists, laboratory technicians, lawyers, and politicians. This coordination of effort is obvious in many areas of conservation biology, and recently has also become a dominant theme in tiger beetle studies (Knisley and Hill 1992; Vogler et al. 1993; Moritz 1994; Vogler and Desalle 1994; Vogler 1998).

Some promising future uses of tiger beetles have direct ramifications for conservation biology, and most of them will involve teams that are interdisciplinary. These areas include climate change (Ashworth 2001), reintroductions (Omland 2002; Brust 2002; Knisley et al. 2005), habitat reclamation (Hussein 2002), habitat management (Omland 2004) and location of conservation reserves and parks (Mittermeier and Mittermeier 1997; Desender and Bosmans 1998; Andriamampianina et al. 2000; Pearson and Carroll 2001; Mittermeier et al. 2004).

Step 6: Technical terminology and methodology so refined they now limit the audience that can fully comprehend it

Although communication with amateurs and the public is a stated goal of the developing cadre of professional conservation biologists, growing reliance on increasingly complex technology and terminology, mathematical models, sophisticated statistics and computer programs have excluded many amateurs and even some professionals in related fields.

For tiger beetles, the rapidly growing use of highly sophisticated disciplines, such as molecular biology, statistical modeling, and satellite imagery have introduced many technical words and concepts. This jargon, in turn, can quickly limit comprehension to a narrow array of associated professionals. As measured in terms of scientific discourse, this trend includes increasing length and number of published articles, increasing sentence complexity, use of multi-word noun phrases, as well as narrowly defined technical terms. It is also well advanced among tiger beetle workers, especially in complex fields, such as molecular studies (Galián et al. 1990; Morgan et al. 2000; Proença and Galián 2003; Goldstein and Desalle 2003; Pons et al. 2004) and mathematical modeling (Carroll and Pearson 2000; Van Dooren and Matthysen 2004).

Paradoxically, although the often-growing complexity of terminology and methodology used in advanced studies of tiger beetles may have excluded most amateurs and many traditional taxonomists, ecologists and behavioral researchers, it appears to have attracted others. For instance, molecular biologists and mathematical modelers seeking appropriate systems on which to apply their technology have used data from tiger beetles with little previous knowledge of the animals themselves. Also, when the U.S. Fish and Wildlife Service listed several tiger beetle species as endangered or threatened, economists, sociologists, foresters, politicians, land owners and members of many unrelated fields, who had little or no previous interest in these taxa, suddenly needed to know about them.

At this point in the march of scientific history, the exclusion of tiger beetle amateurs from complex molecular and statistical studies, while lamentable is not debilitating. However, for conservation biology, just as the field of study reaches a high level of scientific rigor that knowingly will exclude many participants, it simultaneously reaches a point where it must communicate with a growing number of essential participants. Many of these participants are unlikely to comprehend the message or be able to interpret the results of the increasingly complex but more reliable scientific effort. The legislators, judges, lawyers, teachers, and reporters who are critical for implementing policy decisions may not be able to understand the data and generalizations upon which they are basing their decisions. These apparently mutually exclusive goals and effort *are* potentially debilitating.

Discussion

Do the histories of tiger beetle studies and conservation biology follow the model?

Both tiger beetle studies and conservation biology show patterns of change over their history consistent with the GCSPN. However, conservation biology has done so at a velocity that often blurs the progression. Studies of tiger beetles took hundreds of years to arrive at Step 6 and in the last 25 years have become greatly entwined with conservation biology. Conservation biology took less than a century to reach this level, and most of the steps were passed in the last 20 years (Primack 2002).

Although the GCSPN model appears to have broad relevance as shown in its application to the brief history of conservation biology, the rapid advance of this field has apparently obscured some imperfections of the model along the way. Two significant questions need to be answered if the model is to be reliably applied to conservation biology planning.

(1) Are the steps deterministic and inevitable or are they mutable tendencies? Because the major goal of conservation biology is to protect biological diversity while providing for sustainable human needs (Primack 2002), it often seeks to change the outcome of environmental, economic and sociological trends, such as those associated with extinction and habitat destruction. If the general patterns of the GCSPN model represent tendencies that lend themselves to peremptory changes, the model can be used to anticipate problems and implement useful changes to better meet the goals of conservation biology. On the other hand, if the general patterns of the GCSPN model represent inevitable results, the changes fundamental to conservation biology goals are unlikely to be accomplished using these general steps of science development (Myers 1989; Eldredge 1998).

(2) Is each step of the model dependent on the previous step, and if so, how well-developed must a step be before the subsequent step can be initiated and developed? For instance, academic and government support for naming and revising taxa and basic studies of natural history has been in decline for decades and is unlikely to reverse course. As crisis managers, conservation biologists are often forced to make studies on taxa, natural communities and habitats that have severely incomplete foundations of knowledge, such as taxonomy and natural history (Wilson 2000; Hopkins and Freckleton 2002; Dubois 2003; Giangrande 2003). In terms of the GCSPN, the temptation is to yield to the pressures of crisis management and justify a leap from Step 1 to Step 4 or 5 with, perhaps, insufficient investment in the intermediate and supportive steps.

Such a problem evidently occurred with the development of the use of bioindicators in the 1980s and 1990s. Several conservation biologists urged that these surrogate taxa be chosen carefully with predetermined ideal characteristics for a particular use and habitat or ecosystem (Brown 1991; Pearson and Cassola 1992). Unfortunately, many subsequent articles advocating taxa as bioindicators ignored or failed to adequately justify the choice of bioindicators based on predetermined criteria such as reliable taxonomy and basic natural history knowledge. As a result, the credibility of these poorly qualified taxa was challenged, and support of the entire concept of bioindicators quickly diminished (Lawton et al. 1998; Schwartz 1999; Andelman and Fagan 2000; Dale and Beyler 2001).

In the same vein, the U.S. federal Endangered Species Act (ESA) was authorized in 1973. During its tenure, it has engendered considerable controversy, and its future is uncertain (Czech and Krausman 2001). Although property rights, conflicting economic interests, and politics have contributed to many of the controversies, testimony to U.S. congressional committees (Legislative Hearing on H.R. 2829 and H. R. 3705, 20 March 2002) by both conservation advocates and the Assistant Secretary for Fish and Wildlife and Parks place much of the blame for shortcomings of the ESA on poor scientific standards and lack of adequate independent scientific review of endangered species listings. For instance, in one official list of 36 species planned to be delisted in 1999 by then Secretary of the Interior, Bruce Babbitt, five species were already extinct by that time, four were based on taxonomic errors, and ten had been originally listed because of data errors. In this case 53% of these species should not have been on the endangered list in the first place, and a lack of scientific information was to blame (B. Babbitt, pers. com.). A powerful and sophisticated legislative policy assumed that conservation biology was at Step 4 or 5 in the GCSPN, even though Steps 1 and 2 were not sufficiently established to support an advance on to subsequent steps.

What uses does the GCSPN provide for identifying and attaining conservation biology goals?

One important role of the application of the GCSPN model to conservation biology is in providing a context so that we can focus on pertinent questions. At what points should funding agencies support specific efforts? Are there better periods than others in which to attract young recruits to maintain or increase interest in specific taxa or fields such as conservation biology? Can or should dominance by a single individual or small clique be avoided? Will professional biologists exclude the expert amateurs, or will they be able to cooperate?

A second use of the GSCPN is in recognizing broader philosophical problems. For instance, historians of science have shown how cultural differences within national or between regional organizations often dampen paradigm changes in the general area of study (Browne 1996). Can our model illuminate factors such as this and thus avoid intellectual imperialism? Can ideas and hypotheses spread quickly throughout the network, or will resistance to change and other barriers make communication ponderous? Is there a Step 7 in our GCSPN model?

Finally, these preliminary results from comparisons of tiger beetles and conservation biology highlight some specific actions that can be taken immediately. For instance in the area of communication between technical and popular audiences, a basic conservation biology goal, should or can we avoid or ameliorate Step 6?

(1) Three simple changes in writing style and editorial format could make communication easier across a spectrum of readers. First, the abstract and summary of an article can be written in a style that simplifies complex concepts for non-professionals (Gopen and Swan 1990; Knight 2003). Second, for many non-scientific readers, citations in parentheses may become a barrier that disrupts comprehension, a possibility rarely addressed or tested by scientific authors (Rudolph 2003). Using less obtrusive superscript numbers to key citations is one simple change that might broaden communication. This format is already used in several prestigious journals, such as Science, Nature and Trends in Ecology and Evolution. Third, even though metaphors are central to how we think about things, especially when explaining complex concepts to the uninitiated (Short 2000), the editors of some journals, such as Conservation Biology, explicitly discourage authors from using metaphors. Encouraging the use of suitable metaphors to enhance communication might prove more appropriate (Chew and Laubichler 2003).

(2) Although administrators and professional colleagues may demand publications in peer-reviewed journals for promotion and tenure, less prestigious methods for communicating results to the public, such as newspaper and magazine articles, books, and web sites, must receive more than a tacit blessing.

(3) Even though most professional conservation biologists lack the talent or time to communicate with the public as well as Rachel Carson, Jared Diamond, Aldo Leopold, or E.O. Wilson, there are talented science writers, such as David Quamen, Jonathan Weiner and Peter Matthiessen, who can make complex scientific writing comprehensible and attractive to a wide range of the public who have little or no science background. Cooperating with these types of writers, even though credit may be diluted, could disseminate critical information effectively to a wider audience.

(4) Descriptions of new species of tiger beetles, natural history observations, geographical distributions, and seasonal records of occurrence and dispersion, as in many taxa, have by default been turned over largely to expert amateurs. However, not all professionals accept the resultant data as reliable. Recently the British social critics, Charles Leadbeater and Paul Miller (2004), identified a rapidly-growing involvement of amateurs in science from astronomy to medicine that is not fully recognized or utilized. These investigators are a new breed of largely self-trained experts or professional amateurs (Pro-Ams) who, using modern technology, such as the Internet, are producing significant innovations and discoveries in a wide range of fields. Both the government and professionals need to facilitate the contributions of Pro-Ams and be prepared to share the stage with them.

Conclusions

As is typical of model-testing, results often reveal exceptions, unforeseen data, and other anomalies. One accepted procedure in the face of such problems, is to incorporate these unexpected results into a more generalized and useful model. From results of our preliminary consilience tests of the GCSPN, several changes are evident that would make the model more useful. For instance, the history of tiger beetles shows that productive researchers can be working simultaneously in several if not all the steps, especially at later times in the history of a scientific field. Thus it might be more accurate to consider the steps as benchmarks in a continuum rather than linear chronological progressions or irreversible advances. Also, even within well-defined taxa, amateur and professional lines of change appear to diverge into parallel lines rather than follow a single evolving line of science used in the original model (Battalio 1998) (Fig. 1). These parallel lines often have cross lines of influence and varying levels of communication. The different lines each may have their own characteristic benchmarks (Fig. 3). It is also obvious that broader fields, such as conservation biology, build on the work of contributing areas of interest and incorporate their histories rather than follow an independent disciplinary track. Thus, in these succeeding fields, the velocity of change along the time continuum could be expected to be faster and with entire groups of lines converging.

To better understand the model, future tests are needed to clarify not only its patterns but the causes of the patterns. Consilience comparisons of the history of additional taxa or disciplines are one obvious approach. Do all taxa and fields follow the same sequence of steps? Do some histories reveal accelerated progress through certain steps and not others? Accumulated similarities and differences among these histories will provide opportunities to look for their causes. Do factors such as species numbers, their conspicuousness, economic importance, number of investigators, and level of research funding influence patterns and advances in the progression of steps within a field? With an understanding of various combinations of characteristics that might cause differences in development or speed of change, we would be in a better position to understand and apply the model. Among insects, taxa such as ants, cerambycid beetles, scarab beetles, butterflies, dragonflies, and termites would be good candidates for test organisms (New 1984; 1991; 1998; Gaston et al. 1993; Samways 1994; 2005). Comparisons of the history of fields such as wildlife biology, population genetics, and landscape ecology could also be enlightening.

With some immediate solutions and the promise of even more important long range solutions made possible by examinations of historical models, such as the GCSPN, we can be encouraged that conservation biology can make use of its history and learn from it. For instance, Leadbeater and Miller's thesis indicates that with conscious effort the diverging model in Fig. 3 might morph eventually into a model with converging lines, at least between amateurs and early steps in the professional progression of changes. With improvements in the model and future tests of the process of science itself, we may have the best chance to develop foresight, learn from history, and better know if and what changes can be made to better reach our goals. We need not be doomed to repeat history, and even more positively, it may well be that, "We know the future only by the past we project into it" (Gaddis 2004).

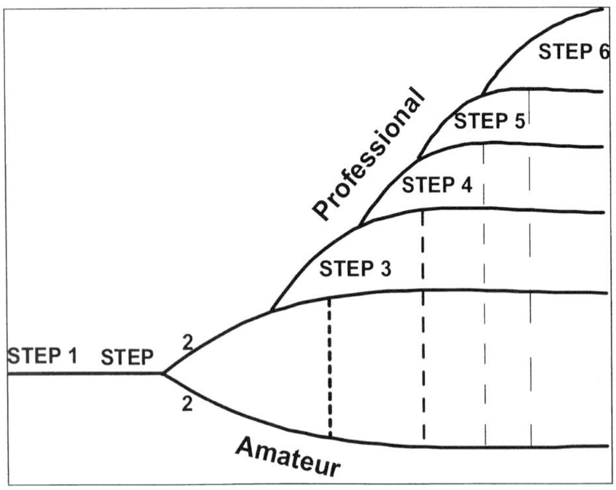

Fig. 3 More complex relation of steps in the GCSPN model as suggested by assessment of the history of tiger beetle studies and conservation biology. Vertical dashed lines indicate varying degrees of communication between amateurs and professionals in the progression

Acknowledgements We are indebted to J. Alcock, K.R. Johnson, C.B. Knisley, B.A. Minteer, N.B. Pearson, S.J. Pyne and D. Sikes for reading early drafts of this manuscript and providing helpful criticism to improve it. J.H. Acorn first challenged us to think about some of these problems and solutions.

References

Acemoglu D, Johnson S, Robinson JA (2001) The colonial origins of comparative development: an empirical investigation. Am Econ Rev 91:1369–1401

Acorn JH (2001) Tiger beetles of Alberta: killers on the clay, stalkers on the sand. Univ. of Alberta Press, Edmonton, Alberta

Andelman SJ, Fagan WF (2000) Umbrellas and flagships: efficient conservation surrogates or expensive mistakes? PNAS 97:5954–5959

Andriamampianina L, Kremen C, Vane-Wright D, Lees D, Razafimahatratra V (2000) Taxic richness patterns and conservation evaluation of Madagascan tiger beetles (Coleoptera: Cicindelidae). J Insect Cons 4:109–128

Arndt E, Aydin N, Aydin G (2005) Tourism impairs tiger beetle (Cicindelidae) populations—a case study in a Mediterranean beach habitat. J Insect Cons 9:201–206

Ashworth AC (2001) Chp. 8: perspectives on quaternary beetles and climate change. In: Gerhard LC, Harrison BM (eds) Geological perspectives of global climate change. American Association of Petroleum Geologists Studies in Geology #47, Tulsa, Oklahoma, pp 153–168

Avise JC (1994) Molecular markers, natural history and evolution. Chapman and Hall, New York

Babione M (2003) Bringing tiger beetles together. Endangered Species Bull 28:28–29

Barraclough TG, Vogler AP (2002) Recent diversification rates in North American tiger beetles estimated from a dated mtDNA phylogenetic tree. Mol Biol Evol 19:1706–1716

Barrow MV Jr (1998) A passion for birds: American ornithology after Audubon. Princeton University Press, Princeton, New Jersey

Battalio JT (1998) The rhetoric of science in the evolution of American ornithological phases. ATTW contemporary studies in technical communication, vol. 8. Ablex Publishing Corporation, Stamford, Connecticut

Bauer KL (1991) Observations on the developmental biology of *Cicindela arenicola* Rumpp (Coleoptera, Cicindelidae). Great Basin Nat 51:226–235

Berglind S-A, Ehnström B, Ljungberg H (1997) Strandskalbaggar, biologisk mångfald och reglering av små vattendrag—exemplen Svartån och Mjällån. Entomologisu Tidskrift 118:137–154

Bossart JL, Carlton CE (2002) Insect conservation in America: status and perspectives. Am Entomol 48:82–92

Brown KS Jr (1991) Conservation of neotropical environments: insects as indicators. In: Collins NM, Thomas JA (eds) Conservation of Insects and their Habitats. Academic Press, London, United Kingdom, pp 349–404

Browne J (1996) Biogeography and empire. In: Jardine N, Secord JA, Spary EC (eds) Cultures of natural history. Cambridge Univ. Press, Cambridge, United Kingdom, pp 305–321

Brust ML (2002) Reintroduction study on *Cicindela formosa generosa* in Marinette County, Wisconsin. Cicindela 34:5–7

Carroll SS (1998) Modelling abiotic indicators when obtaining spatial predictions of species richness. Environ Ecol Stat 5:257–276

Carroll SS, Pearson DL (1998a) Spatial modeling of butterfly species richness using tiger beetles (Cicindelidae) as a bioindicator taxon. Ecol Appl 8:531–543

Carroll SS, Pearson DL (1998b) The effects of scale and sample size on the accuracy of spatial predictions of tiger beetle (Cicindelidae) species richness. Ecography 21:401–414

Carroll SS, Pearson DL (2000) Detecting and modeling spatial and temporal dependence in conservation biology. Cons Biol 14:1893–1897

Carter M (1990) The idea of expertise: an exploration of cognitive and social dimensions of writing. College Compos Commun 41:265–286

Cassola F (2002) Le cicindele e le coste: biogeografia e conservazione (Studi sui Cicindelidi. CXIX). Biogeographia 23:55–69

Cassola F, Pearson DL (2000) Global patterns of tiger beetle species richness (Coleoptera: Cicindelidae): their use in conservation planning. Biol Cons 95:197–208

Chafe W (1986) Evidentiality in English conversation and academic writing. In: Chafe W, Nichols J (eds) Evidentiality: the linguistic coding of epistemology. Albex Press, Norwood, NJ, pp 261–272

Chew MK, Laubichler MD (2003) Natural enemies: metaphor or misconception? Science 301:52–53

Choate PM (2003) Tiger beetles: a field guide and identification manual for Florida and Eastern US. Univ. Press of Florida, Gainesville, Florida

Crandall KA, Bininda-Emonds ORP, Mace GM, Wayne RK (2000) Considering evolutionary processes in conservation biology. Trends Ecol Evol 15:290–295

Cressie N (1991) Statistics for spatial data. John Wiley, New York, NY

Czech B, Krausman PR (2001) The endangered species act: history, conservation biology, and public policy. Johns Hopkins Univ. Press, Baltimore, Maryland

Dale VH, Beyler SC (2001) Challenges in the development and use of ecological indicators. Ecol Indicators 1:3–10

Desender K, Bosmans R (1998) Ground beetles (Coleoptera, Carabidae) on set-aside fields in the Campine region and their importance for nature conservation in Flanders (Belgium). Biodivers Conserv 7:1485–1493

Desender K, Dufrêne M, Maelfait J-P (1994) Long term dynamics of carabid beetles in Belgium: a preliminary analysis on the influence of changing climate and land use by means of a data base covering more than a century. In: Desender K et al. (eds) Carabid beetles: ecology and evolution. Kulwer Academic Publishers, Netherlands, pp 247–252

Desender K, Turin H (1989) Loss of habitats and changes in the composition of the ground and tiger beetle fauna in four west European countries since 1950 (Coleoptera: Carabidae, Cicindelidae). Biol Cons 48:277–294

Diamond J (2002) Evolution, consequences and future of plant and animal domestication. Nature 418:700–707

Diogo AC, Vogler AP, Gimenez A, Gallego D, Galián J (1999) Conservation genetics of *Cicindela deserticoloides*, an endangered tiger beetle endemic to southeastern Spain. J Insect Cons 3:117–123

Dreisig H (1980) Daily activity, thermoregulation and water loss in the tiger beetle *Cicindela hybrida*. Oecologia 44:376–389

Dubois A (2003) The relationships between taxonomy and conservation biology in the century of extinctions. Comptes Rendus Biologies 326(supplement):9–21

Eldredge N (1998) Life in the balance. Humanity and the biodiversity crisis. Princeton Univ. Press, Princeton, NJ

Farber PL (2000) Finding order in nature: The naturalist's tradition from Linnaeus to E.O. Wilson. Johns Hopkins University Press, Baltimore Maryland

Gaddis JL (2004) The landscape of history: how historians map the past. Oxford Univ. Press, New York, New York

Galián J, Vogler AP (2003) Evolutionary dynamics of a satellite DNA in the tiger beetle species pair *Cicindela campestris* and *C. maroccana*. Genome 46:213–223

Galián J, Serrano J, Ortiz AS (1990) Karyotypes of nine species of Cicindelini and cytotaxonomic notes on Cicindelinae (Coleoptera, Carabidae). Genetica 82:17–24

Gaston KJ, New TR, Samways MJ (1993) Perspectives on insect conservation. Intercept Publ., Andover, Hants, UK

Giangrande A (2003) Biodiversity, conservation, and the "taxonomic impediment." Aquat Conserv-Marine Freshwater Ecosys 13:451–459

Gilbert C (1997) Visual control of cursorial prey pursuit by tiger beetles (Cicindelidae). J Comp Physiol A 181:217–230

Goldstein PZ, Desalle R (2003) Calibrating phylogenetic species formation in a threatened insect using DNA from historical specimens. Mol Ecol 12:1993–1998

Goldstein PZ, Desalle R, Amato G, Vogler AP (2000) Conservation genetics at the species boundary. Cons Biol 14:120–131

Gopen GD, Swan JA (1990) The science of scientific writing. Am Sci 78:550–558

Gould SJ (1989) Wonderful life: the burgess shale and the nature of history. Norton, New York, New York

Greenberg CH, McGrane A (1996) A comparison of relative abundance and biomass of ground-dwelling arthropods under different forest management practices. Forest Ecol Manag 89:31–41

Hadley NF, Schultz TD, Savill A (1988) Spectral reflectances of three tiger beetle subspecies (*Neocicindela perhispida*): correlation with their habitat substrate. NZ J Zool 15: 343–346

Hibbs DA, Olsson O (2004) Geography, biogeography, and why some countries are rich and others are poor. PNAS USA 101:3715–3720

Hoback WW, Podrabsky JE, Higley LG, Stanley DW, Hand SC (2000) Anoxia tolerance of con-familial tiger beetle larvae is associated with differences in energy flow and anaerobiosis. J Comp Physiol B 170:307–314

Holeski PM, Graves RC (1978) An analysis of the shore beetle communities of some channelized streams in northwest Ohio (Coleoptera). Great Lakes Entomol 11:23–36

Hopkins GW, Freckleton RP (2002) Declines in the numbers of amateur and professional taxonomists: implications for conservation. Anim Conserv 5:245–249

Horgan FG, Chávez JC (2004) Field boundaries restrict dispersal of a tropical tiger beetle, *Megacephala angustata* Chevrolet 1841 (Coleoptera: Cicindelidae). Entomotropica 19:147–152

Hussein ML-a (2002) Ground beetle (Coleoptera, Carabidae) communities of the exhausted open-cast mining area K™nigsaue and of agricultural areas in the District Aschersleben-sta§furt (Saxony-Anhalt). Arch Phytopathol Plant Protect 35:125–155

Kamoun S (1996) Occurrence of the threatened *Cicindela senilis frosti* Varas Arangua in an inland salt marsh in Riverside County, California (Coleoptera: Cicindelidae). Coleop Bull 50:369–371

Killingsworth MJ, Palmer JS (1992) Ecospeak: rhetoric and environmental politics in America. So. Illinois Univ. Press, Carbondale, Illinois

Kitching IJ (1996) Identifying complementary areas for conservation in Thailand: an example using owls, hawkmoths and tiger beetles. Biodivers Conserv 5:841–858

Knight J (2003) Scientific literacy: clear as mud. Nature 423: 376–378

Knisley CB, Fenster MS (2005) Apparent extinction of the tiger beetle, *Cicindela hirticollis abrupta* (Coleoptera: Carabidae: Cicindelinae). Coleop Bull 59:451–458

Knisley CB, Hill JM (1992) Effects of habitat change from ecological succession and human impact on tiger beetles. Virginia J Sci 43:134–142

Knisley CB, Hill JM (2001) Biology and conservation of the coral pink sand dunes tiger beetle, *Cicindela limbata albissima*. West N Am Nat 61:381–394

Knisley CB, Hill JM, Scherer AM (2005) Translocation of threatened tiger beetle *Cicindela dorsalis dorsalis* (Coleoptera: Cicindelidae) to Sandy Hook, New Jersey. Ann Ent Soc Am 98:552–557

Knisley CB, Schultz TD (1997) The biology of tiger beetles and a guide to the species of the South Atlantic States. Virginia Mus Nat Hist, Special Publication No. 5, Martinsville, Virginia

Kremen C (1992) Assessing the indicator properties of species assemblages for natural areas monitoring. Ecol Appl 2: 203–217

Kremen C, Colwell RK, Erwin TL, Murphy DD, Noss RF, Sanjayan MA (1993) Terrestrial arthropod assemblages: their use as indicators in conservation planning. Cons Biol 7:796–808

Lakoff G, Johnson M (1980) Metaphors we live by. Univ. Chicago Press, Chicago, Illinois

Lawton JH, Bignell DE, Bolton B, Bloemers GF, Eggleton P, Hammond PM, Hodd M, Holt RD, Larsen TB, Mawdsley NA, Stork NE, Srivastava DS, Watt AD (1998) Biodiversity in inventories, indicator taxa and effects of habitat modification in tropical forest. Nature 391:72–76

Leadbeater C, Miller P (2004) The pro-Am revolution: how enthusiasts are changing our economy and society. Demos Publ., London, United Kingdom

Lees DC, Kremen C, Andriamampianina L (1999) A null model for species richness gradients: bounded range overlap of butterflies and other rainforest endemics in Madagascar. Biol J Linn Soc 67:529–584

Leonard JG, Bell RT (1999) Northeastern tiger beetles: a field guide to tiger beetles of New England and Eastern Canada. CRC Press, Boca Raton, Florida

Linné Kv (1758) Systema naturae, ed. 10 (Systema naturae per regna tria naturae, secundum classes, ordines, genera, species, cum characteribus, differentiis, synonymis, locis. Tomus I. Editio decima, reformata). Holmiae

Maguire LA (1991) Risk analysis for conservation biologists. Cons Biol 5:123–125

Mawdsley JR (2005) Extirpation of a population of *Cicindela patruela* Dejean (Coleoptera: Carabidae: Cicindelini) in suburban Washington, DC, USA. Proc Entomol Soc Wash 107:64–70

Maienschein J (2000) Why study history for science? Biol Philos 15:339–348

McGeoch MA (1998) The selection, testing and application of terrestrial insects as bioindicators. Biol Rev 73:181–201

Meine C (1999) It's about time: conservation biology and history. Cons Biol 13:1–3

Mittermeier RA, Gil PR, Hoffmann M, Pilgrim J, Brooks T, Mittermeier CG, Lamoreux J, Da Fonseca G (2004) Hotspots revisted: earth's biologically richest and most endangered terrestrial ecosystems. CEMEX, Mexico City, Mexico

Mittermeier R, Mittermeier CG (1997) Megadiversity: earth's biologically wealthiest nations. CEMEX, Mexico City, Mexico

Morgan M, Knisley CB, Vogler AP (2000) New taxonomic status of the endangered tiger beetle *Cicindela limbata albissima* (Coleoptera: Cicindelidae): evidence from mtDNA. Ann Entomol Soc Am 93:1108–1115

Moritz C, (1994) Defining "Evolutionary Significant Units" for conservation. Trends Ecol Evol 9:373–375

Murphy DD, Freas KE, Weiss SB (1990) An environment-metapopulation approach to population viability analysis for a threatened invertebrate. Cons Biol 4:41–51

Myers N (1989) A major extinction spasm: predictable and inevitable? In: Western D, Pearl MC (eds) Conservation for the twenty-first century. Oxford Univ. Press, Oxford, United Kingdom, pp 42–49

Nagano CD (1980) Population status of the tiger beetles of the genus *Cicindela* (Coleoptera: Cicindelidae) inhabiting the marine shoreline of Southern California. Atala 8:33–42

New TR (1984) Insect conservation: an australian perspective. Dordrecht Publ., Hingham, Massachusetts

New TR (1991) Butterfly conservation. Oxford Univ. Press, Oxford, UK

New TR (1998) Invertebrate surveys for conservation. Oxford Univ. Press, Oxford, UK

Noss RF (1990) Indicators for monitoring biodiversity: a hierarchical approach. Cons Biol 4:355–364

Nye MJ (1996) Before big science: the pursuit of modern chemistry and physics, 1800–1940. Twayne Publishers, New York, New York

Okamura J-Y, Toh Y (2004) Morphological and physiological identification of medulla interneurons in the visual system of the tiger beetle larva. J Comp Physiol A 190:449–468

Omland KS (2002) Larval habitat and reintroduction site selection for *Cicindela puritana* in Connecticut. Northeastern Nat 9:433–450

Omland KS (2004) Puritan tiger beetle (*Cicindela puritana*) on the Connecticut River: habitat management and translocation alternatives. In: Akçakaya HR, Burgman MA, Kindvall O, Wood CC, Sjögren-Gulve P, Hatfield JS, McCarthy MA (eds) Species conservation and management: case studies. Oxford Univ. Press, London, pp 137–148

Pearson DL (1994) Selecting indicator taxa for the quantitative assessment of biodiversity. Philos Trans R Soc Lond B 345:75–79

Pearson DL (1988) Biology of tiger beetles. Ann Rev Entomol 33:123–147

Pearson DL (2004) A list of suggested common English names for species of tiger beetles (Coleoptera: Cicindelidae) occurring in Canada and the US. Cicindela 36:31–39

Pearson DL, Carroll SS (1998) Global patterns of species richness: spatial models for conservation planning using bioindicators and precipitation. Cons Biol 12:809–821
Pearson DL, Carroll SS (1999) The influence of spatial scale on cross-taxon congruence patterns and prediction accuracy of species richness. J Biogeog 26:1079–1090
Pearson DL, Carroll SS (2001) Predicting patterns of tiger beetle (Coleoptera: Cicindelidae) species richness in northwestern South America. Stud Neotrop Fauna Environ 36:123–134
Pearson DL, Cassola F (1992) World-wide species richness patterns of tiger beetles (Coleoptera: Cicindelidae): indicator taxon for biodiversity and conservation studies. Cons Biol 6:376–391
Pearson DL, Cassola F (2005) A quantitative analysis of species descriptions of tiger beetles (Coleoptera: Cicindelidae), from 1758 to 2004, and notes about related developments in biodiversity studies. Coleop Bull 59:184–193
Pearson DL, Knisley CB, Kazilek CJ (2005) A field guide to the tiger beetles of the United States and Canada: identification, natural history and distribution of the cicindelids. Oxford Univ. Press, New York, New York
Pearson DL, Shetterly JA (2006) How do published field guides influence interactions between amateurs and professionals in entomology? Am Entomol 52. In press
Pearson DL, Vogler AP (2001) Tiger beetles: the evolution, ecology and diversity of the cicindelids. Cornell Univ. Press, Ithaca, New York
Pons J, Barraclough TG, Theodorides K, Cardoso A, Vogler AP (2004) Using exon and intron sequences of the gene Mp20 to resolve basal relationships in *Cicindela* (Coleoptera: Cicindelidae). Syst Biol 53:554–570
Primack RB (2002) Essentials of conservation biology, 3rd edn. Sinauer Associates, Sunderland, Massachusetts
Proença SJR, Galián J (2003) Chromosome evolution in the genus *Cicindela*: physical mapping and activity of rDNA loci in the tiger beetle species *Cicindela littoralis* and *C. flexuosa*. J Zool Syst Evol Res 41:227–232
Proença SJR, Serrano ARM, Collares-Pereira J (2002) An unusual karyotype with low chromosome number in Megacephalini, a basal group of tiger beetles (Coleoptera, Cicindelidae): cytogenetic characterisation by C-banding and location of rDNA genes. Hereditas 137:202–207
Pyle R, Bentzien M, Opler P (1981) Insect conservation. Ann Rev Entomol 26:233–258
Richoux P (2001) Sensibilité de *Cylindera arenaria* aux aménagements fluviaux: l'example de la région lyonnaise (Coléptères Cicindelidae). Cahiers scientifique Mus.d'Histoire nat. Lyon 2:63–74
Ricketts TH, Dinerstein E, Olson DM, Loucks C (1999) Who's where in North America? Patterns of species richness and the utility of indicator taxa for conservation. Bioscience 49:369–381
Rivers-Moore NA, Samways MJ (1996) Game and cattle tramping, and impacts of human dwellings on arthropods at a game park boundary. Biodivers Conserv 5:1545–1556
Rodríguez JP, Pearson DL, Barrera RR (1998) A test for the adequacy of bioindicator taxa: are tiger beetles (Coleoptera: Cicindelidae) appropriate indicators for monitoring the degradation of tropical forests in Venezuela? Biol Cons 83:69–76
Rolett B, Diamond J (2004) Environmental predictors of pre-European deforestation on Pacific Islands. Nature 431: 443–446
Rudolph VJ (2003) Modifying forest management for biodiversity. Cons Biol 17:1463–1571
Samways MJ (1994) Insect conservation biology. Chapman and Hall, London, UK
Samways MJ (2005) Insect diversity conservation. Cambridge Univ. Press, Cambridge, UK
Satoh A, Sota T, Uéda T, Enokido Y, Paik§ C, Hori M (2004) Evolutionary history of coastal tiger beetles in Japan based on a comparative phylogeography of four species. Mol Ecol 13:3057–3069
Schultz TD (1988) Destructive effects of off-road vehicles on tiger beetle habitat in central Arizona. Cicindela 20:25–29
Schwartz MW (1999) Choosing the appropriate scale of reserves for conservation. Ann Rev Ecol Syst 30:83–108
Shelford R (1902) Observations on some mimetic insects and spiders from Borneo and Singapore. Proc Zool Soc London 1902:230–284
Shelford VE (1907) Preliminary note on the distribution of the tiger beetles (*Cicindela*) and its relation to plant succession. Biol Bull Marine Biol Lab, Woods Hole 14:9–14
Short DC (2000) Analysing metaphor in human resource development. Human Resour Devel Int 3:323–341
Siemann W (eds) (2003) Umweltgeschichte: Themen und Perspektiven. Verlag C. H. Beck oHG, Munich, Germany
Sikes DS (2002) Beetles of Block Island: rare species that once occurred on the mainland, Chapter 11. In: Paton PW, Gould LL, August PV, Frost AO (eds) The ecology of Block Island: proceedings of the Rhode Island natural history survey conference, October 28, 2000. Rhode Island Natural History Survey, Kingston, RI, pp 183–191
Sorensen WC (1995) Brethren of the net: American entomology 1840–1880. Univ. of Alabama Press, Tuscaloosa, Alabama
Trautner J (1996) Historische und aktuelle Bestandssituation des Sandlaufkäfers *Cicindela arenaria* Fuesslin, 1775 in Deutschland (Col., Cicindelidae). Entomol Nact Berichte 40: 83–88
USDI (1980) Habitat evaluation procedures (HEP). Ecological services manual number 102. Division of Ecological Services, USDI. Fish and Wildlife Service, Washington, DC
Van Dooren TJM, Matthysen E (2004) Generalized linear models for means and variances applied to movements of tiger beetles along corridor roads. J Anim Ecol 73: 261–271
Vogler AP, Knisley CB, Glueck SB, Hill JM, Desalle R (1993) Using molecular and ecological data to diagnose endangered populations of the Puritan Tiger Beetle *Cicindela puritana*. Mol Ecol 2:375–383
Vogler AP (1998) Extinction and the evolutionary process in endangered species: what to conserve? In: DeSalle R, Schierwater B (eds) Molecular approaches to ecology and evolution. Birkhäuser Verlag, Basel, Switzerland, pp 191–210
Vogler AP, DeSalle R (1994) Diagnosing units of conservation management. Cons Biol 8:354–363
Wickham HF (1904) The influence of the mutations of the Pleistocene lakes upon the present distribution of *Cicindela*. Am. Nat. 38:643–654
Wilson EO (2000) On the future of conservation biology. Cons Biol 14:1–3
Yarbrough WW, Knisley CB (1994) Distribution and abundance of the coastal tiger beetle, *Cicindela dorsalis media* (Coleoptera: Cicindelidae), in South Carolina. Ent News 105: 189–194
Zirnstein G (1996) Ökologie und Umwelt in der Geschichte. Ökologie und Wirtschaftforschung, Band 14. Metropolis-Verlag, Marburg, Germany

J Insect Conserv (2007) 11:61–73
DOI 10.1007/s10841-006-9019-8

BEETLE CONSERVATION

Unfortunate encounters? Novel interactions of native *Mecyclothorax*, alien *Trechus obtusus* (Coleoptera: Carabidae), and Argentine ant (*Linepithema humile*, Hymenoptera: Formicidae) across a Hawaiian landscape

James K. Liebherr · Paul D. Krushelnycky

Received: 19 November 2005 / Accepted: 1 March 2006 / Published online: 2 November 2006

Abstract The Hawaiian Islands support a speciose radiation of native *Mecyclothorax* beetles (Coleoptera: Carabidae). This lineage has undergone a classical island radiation resulting in extensive ecological specialization, flight-wing loss, and 100% single-island endemism. We report on the sympatric occurrence of several *Mecyclothorax* species endemic to Haleakala volcano, East Maui with the newly arrived, adventive *Trechus obtusus* (Coleoptera: Carabidae), a tramp species originally from Europe. Arrival of *T. obtusus* in afforested, non-native gymnosperm plantation forest near Polipoli, Maui was associated with subsequent decreased abundance of native *Mecyclothorax* beetles. Since discovery of *T. obtusus* on Haleakala, their populations have been transformed through subsequent increase in frequency of brachypterous individuals. Consequences of this transformation to flight-wing dimorphic populations may simultaneously include enhanced reproductive capacity of brachypterous individuals, increased local adaptation of populations, and enhanced metapopulational dynamics ultimately permitting range expansion and occupation far beyond anything observed for the monomorphically brachypterous native *Mecyclothorax*. *Trechus obtusus* and several *Mecyclothorax* species occur sympatrically with Argentine ant, *Linepithema humile* (Hymenoptera: Formicidae) in subalpine shrublands on Haleakala. Recent sampling corroborates earlier findings that localized presence of Argentine ant is associated with significantly decreased abundance of native *Mecyclothorax*. Conversely, abundance of the continental *T. obtusus* is not significantly affected by ant presence.

Keywords Adaptation · Adventive species · Afforestation · Area of endemism · Brachyptery · Invasion

J. K. Liebherr (✉)
Department of Entomology, John H. and Anna B. Comstock Hall, Cornell University, Ithaca, New York 14853-2601, USA
e-mail: JKL5@cornell.edu

P. D. Krushelnycky
Division of Organisms and Environment, Department of Environmental Science, Policy & Management, University of California, 137 Mulford Hall, Berkeley, CA 94720-3114, USA

Introduction

The Hawaiian Islands, with a relatively small terrestrial area of 16,640 km^2, are home to 5138 described native endemic insect species (Eldredge and Evenhuis 2003). These endemic species constitute 63% of the total Hawaiian insect fauna, with the balance consisting of 163 native indigenous species also found elsewhere in the Pacific Ocean region, plus an ever increasing number of nonindigenous species introduced through the activities of man. From 1995 to 2003, 639 nonindigenous species were added to the known State fauna (Eldredge and Evenhuis 2003); an average of 80 species per year. This recent rate of alien introduction is four times the rate observed for the period 1962–1976 (Beardsley 1979). Adverse effects on the native arthropod biota have been documented for a variety of alien social Hymenoptera, including ants (Loope et al. 1988; Cole et al. 1992; Gillespie and Reimer 1993; Reimer 1994, LaPolla et al. 2000; Krushelnycky et al. 2005) and vespid wasps (Gambino et al. 1987;

Gambino 1992). Predaceous foraging activities of these social animals directly impacts populations of native insects, decreasing their population levels within areas infested by these invasive species. Other adverse interactions may also occur. These include interference competition at floral resources between the alien Argentine ant, *Linepithema humile* (Mayr), and native *Hylaeus* bees (Colletidae) (Lach 2004), and parasitism by alien hymenopterous parasitoids, both intentionally and accidentally introduced, on native lepidopterous caterpillars leading to as much as 20% mortality in the larval stages of the hosts (Henneman and Memmott 2001).

The native insect fauna, and indeed the entire native Hawaiian biota, is characterized by high levels of archipelago specific endemism, but moreover by exceedingly restricted geographic ranges within the archipelago, with many species geographically limited to single islands, volcanoes, or even particular flanks of volcanoes. Risk for endangerment is highly associated with geographically restricted ranges among the Hawaiian flora (Sakai et al. 2002). Such geographical restrictions also apply to Hawaiian birds and mollusks (Hadfield et al. 1993), with all three groups contributing to consistently high rankings for Hawaiian counties with regard to endangered species richness (Dobson et al. 1997; Rutledge et al. 2001). Knowledge of the distributions of native Hawaiian insects is more fragmentary, though the association of endangerment with geographic restriction holds for instances where information is available, e.g. *Megalagrion* damselfies (Coenagrionidae) (Polhemus and Asquith 1996; Englund 1999), crickets (LaPolla et al. 2000) and carabid beetles (Liebherr 2004). In the latter instance, extensive logging as well as degradation of native koa forest through grazing of cattle extensively reduced that particular habitat during the past century (Pratt and Gon 1998), leading to serious reduction and possible extinction of geographically restricted, koa forest carabid beetle specialists (Liebherr 2004).

In this paper we examine the interactions among native Hawaiian carabid beetles of the genus *Mecyclothorax* Sharp (Liebherr 2005), an alien invasive carabid beetle *Trechus obtusus* Erichson, and the invasive Argentine ant. We focus on sites on Haleakala volcano, Maui Island where the invasive carabid beetle has become established. Hawaii is home to a speciose radiation of *Mecyclothorax*, with over 90 species currently described (Britton 1948; Liebherr and Zimmerman 2000; Liebherr 2005). All Hawaiian *Mecyclothorax* individuals are brachypterous, and the species are highly endemic in distribution, with none distributed on more than one island. Of the *Mecyclothorax* species investigated in this study, four are known only from an afforested site at Polipoli Springs State Recreation Area, thereby defining this limited extent of the volcano as an area of endemism. The native beetles are found in native forest remnants at Polipoli surrounded by plantation-planted, nonindigenous gymnosperms (Liebherr 2005). Fortuitously, native carabid beetles had been surveyed at this site prior to the discovery of the alien *Trechus obtusus*. Upon and since the arrival of *T. obtusus* at Polipoli, a single site has been revisited at biennial intervals to assess any temporal changes in the relative abundances of the invasive versus the native species.

We also investigate changes that are occurring in the populations of the alien *T. obtusus*, as it adapts to its novel habitat. This adaptation is represented by the origination and subsequent increase in frequency of brachypterous, flightless individuals. In its native European range, *T. obtusus* is known to colonize novel habitats establishing initial populations comprising entirely flight-capable individuals. Populations then quickly transform so that they constitute nearly 100% brachypterous individuals (Den Boer 1968, Desender et al. 1980a, b). We report a similar transformation at sites on Haleakala.

Thirdly, we investigate the interactions among native *Mecyclothorax* species, the alien carabid *T. obtusus*, and the Argentine ant. Previously, populations of the Argentine ant on Haleakala have been shown to be negatively associated with *Mecyclothorax cordithorax* Liebherr (reported as *M. robustus* Sharp in Loope et al. 1988; Cole et al. 1992) as well as the geographically restricted, high elevation specialist carabid beetle species, *Blackburnia lenta* Sharp (Krushelnycky et al. 2005). Such a negative association is not observed between the Argentine ant and *T. obtusus*, suggesting that *T. obtusus* can persist in areas disturbed by the alien ants.

Materials and methods

Study area and sampling

The study area consisted of sites where the adventive alien species, *Trechus obtusus* (Liebherr and Takumi 2002) has become established on East Maui Island, Hawaii (Fig. 1). The species was first discovered in 1999 at the Haleakala National Park Service Area (20°46.28′ N, 156°14.86′ W, el. 2072 m) near the Park entrance at Pu'u Nianiau. This area supports mesic, open shrubland and alien stands of non-native Monterey Pine (*Pinus radiata*). The shrubland habitat is

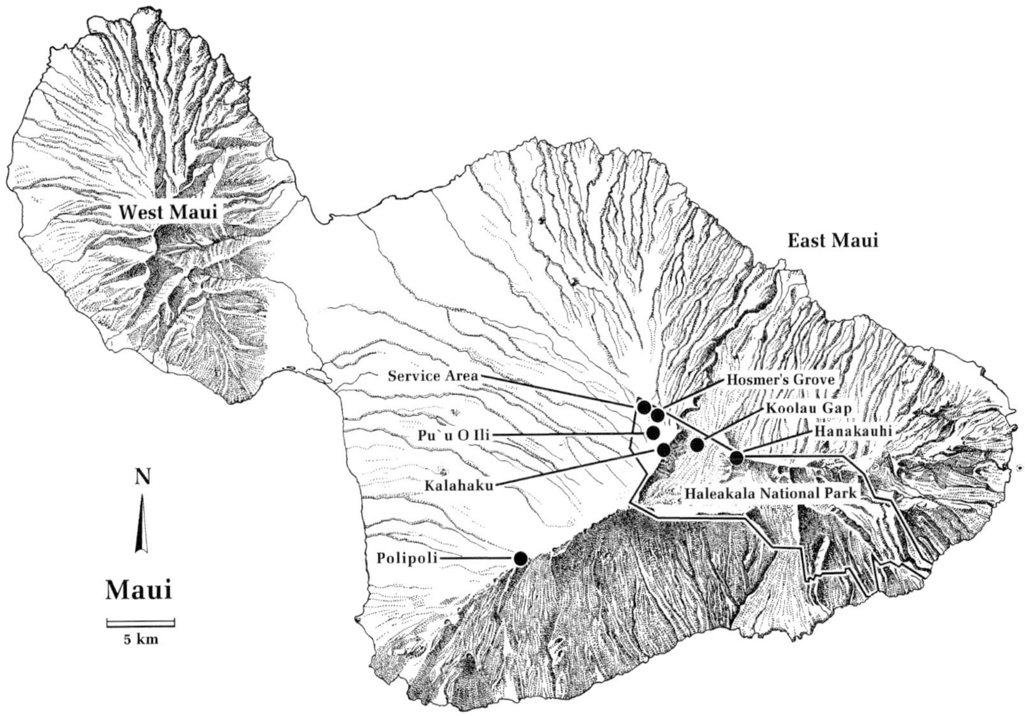

Fig. 1 Maui Island, Hawaii, with sites mentioned in text indicated. Named sites represent all presently known collecting localities for *Trechus obtusus*

dominated by *Sophora chrysophylla* and *Deschampsia nubigena* bunchgrass. In 2001 *T. obtusus* was found inhabiting afforested gymnosperm dominated forest habitats on the southwest rift zone of the leeward face of Haleakala volcano at Polipoli Springs State Recreation Area (20°40.67′ N, 156°20.18′ W, el. 1738–1890 m) (Liebherr and Takumi 2002). *Trechus obtusus* was first found at intervening shrubland sites at Pu'u O Ili (20°44.03′ N, 156°14.25′ W, el. 2360 m) and Kalahaku (20°44.03′ N, 156°14.25′ W, el. 2844 m) in 2002, during studies that assessed Argentine ant populations and their impact on native Hawaiian arthropods (Krushelnycky et al. 2004, 2005). Natural areas to the north and east of the Service Area were surveyed for *T. obtusus* (Liebherr and Takumi 2002, unpubl. data 2003), with no evidence at that time that *T. obtusus* had entered native habitats of Koa (*Acacia koa*) or Ohia lehua (*Metrosideros polymorpha*) dominated forests in the Waikamoi Preserve of The Nature Conservancy of Hawaii that lie down slope and to the north northeast of the National Park. However, a single specimen was found in 2003 along the National Park boundary fence in disturbed, afforested gymnosperm forest habitats near Hosmer's Grove campground (20°46.13′ N, 156°14.12′ W, el. 2130 m), and most recently in November 2005, *T. obtusus* was discovered in Koolau Gap (20°45.09′ N, 156°12.42′ W, el. 1990 m) and to the east on Hanakauhi peak (20°44.18′ N, 156°10.73′ W, el. 2710 m) (Krushelnycky, unpubl. data) (Fig. 1).

Native and introduced carabid beetles (Coleoptera: Carabidae) were sampled at various sites in the study area using a variety of methods. In the non-forest habitats near the Service Area, pitfall trapping was utilized to collect ground-level active adult beetles, and plastic diphacinone rat bait stations were opened at intervals, with all beetles inside removed for study (Liebherr and Takumi 2002). The most extensive samples were obtained from January to March 2002. At Polipoli Springs, beetles were collected by hand through grubbing, whereby litter on the soil surface is pushed aside, with beetles running across the bared soil surface as it dried. Beetles were also found under bark of downed decaying logs, and under the logs on the ground. Logs with mossy growth were sprayed with synthetic pyrethrin insecticide, causing beetles and other insects to exit the log and fall onto sheets placed below.

Quantitative sampling was undertaken near Polipoli Springs, Kula Forest Reserve, at a single site in a mesic, steep-sided gulch (20°40.62′ N 156°20.18′ W, el. 1781 m) characterized by an alien overstory of Arizona cypress (*Cupressus arizonica*) (Little and Skolmen 1989) and a native understory consisting of *Dryopteris wallachiana* ferns. Lone individuals of Kawau or Hawaiian Holly

(*Ilex anomala*) and Kolea (*Myrsine lessertiana*) also occurred in the gulch sample site. The gulch is approximately 10 m wide at the bottom, with sides rising increasingly steeply to flanking ridges about 30–40 m above the bottom. Sampling was conducted on a gulch reach extending approximately 40–50 m from a steep headwall to a rocky step-like outlet that borders plantings of tropical ash (*Fraxinus hydei*). The gulch bottom is covered with a 3–10 cm layer of *Dryopteris* fronds, cypress needles and leaves of tropical ash blown into the gulch. Beetles in the litter were sampled first on 20 May 2001 – the year *Trechus obtusus* was discovered at this site – and then subsequently on 19 May 2003 and 13 May 2005 (Table 1). On each sampling date, leaf litter from square meter quadrats was raked into a Reitter sifter, with the resulting siftate examined for beetles after spreading it on a beating sheet. In 2001, four quadrats – i.e. 4 m^2 of forest floor litter – were examined and pooled. On the two subsequent visits, specimens from individual siftate loads were processed separately. Five m^2 samples were taken in 2003, and seven were gathered and examined for beetles in 2005. In the latter two years, a degree of randomness and independence from previous quadrat sampling was attempted by throwing the beating sheet after turning with eyes closed on a flat shelf along the gulch bottom. The sample was then taken at the flattest square meter touching the sheet. An eighth "taxonomist's" sample was also taken in 2005, in which previously unsampled, non-contiguous moist areas adjacent to *Dryopteris* ferns were sifted; the total siftate equivalent to that of a single m^2 portion of forest floor. This was the only non-contiguous sample taken in the gulch. All siftate was replaced on the bare, raked sample sites, reestablishing the continuous litter layer. Given the total number of m^2 samples, at most 2% of the available gulch floor was disturbed by raking and sifting litter during each biennial sampling date. A χ^2-test (8 degrees of freedom) was used to test for significant differences in total numbers of individuals for the five carabid species at the gulch site, across the three sampling years.

Sampling at Pu'u O Ili and Kalahaku, Haleakala National Park was undertaken using pitfall trapping in the shrubland habitat with trap design, placement and collecting methods described in Krushelnycky et al. (2005). Traps were opened for two weeks in September–October 2002 and July 2003. Documentation of the impact of the Argentine ant on populations of native insects comprised the primary focus of sampling at these sites. As such, pitfall traps were arrayed in eight pairs of matched plots for each of the two areas of ant infestation, with one plot in each pair situated inside the ant range and the mate situated outside the ant range (Fig. 2). Additional plots were established inside the ant ranges for other purposes, and carabid beetle captures from these additional plots (not shown on Fig. 2) were included for calculations of wing morphology frequency (Table 2) but were not included for analyses of relative beetle abundances in the presence and absence of ants. Three pitfall traps were installed in each plot, and catches were pooled for each plot. Relative abundance data for *T. obtusus* and *Mecyclothorax* (all species pooled, including *M. cordithorax*, *M. montivagus* (Blackburn) and *M. apicalis* Sharp) were analyzed with General Linear Models (GLM) in Minitab 14 (Ryan et al. 2005). The models included the fixed factors *Argentine ant presence* (ants, no ants) and

Table 1 Frequencies of *Trechus obtusus* and native *Mecyclothorax* species in Reitter sift samples collected during May at gulch site, Polipoli State Recreation Area, Kula Forest Reserve, Maui

Year and sample	*T. obtusus*	*M. cordithorax*	*M. consobrinus*	*M. irregularis*	*M. aeneipennis*
2001					
(Total of 4 m^2 samples)	77(9)	15	3	2	0
2003					
lot 01	12(1)	0	0	0	0
lot 02	8(0)	0	0	0	0
lot 03	6(2)	0	1	0	0
lot 04	6(2)	1	0	0	0
lot 05	14(3)	1	0	0	0
(Total)	46(8)	2	1	0	0
2005					
lot 01	7(0)	0	1	0	1
lot 02	5(0)	0	2	0	0
lot 03	7(3)	1	0	0	0
lot 04	13(2)	0	0	0	0
lot 05	10(1)	0	0	2	0
lot 06	10(3)	0	0	1	0
lot 07	21(5)	1	2	0	0
(Total)	73(15)	2	5	3	1
"taxonomist's" lot	45(12)	2	1	0	0

Numbers of teneral *T. obtusus* individuals indicated in parentheses. See text for descriptions of samples

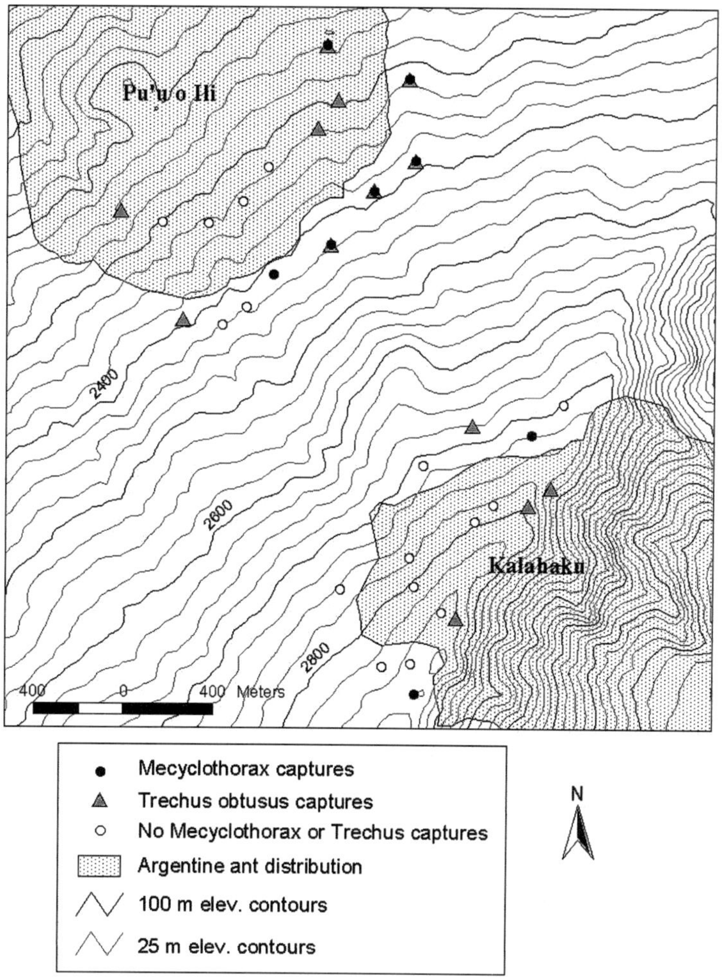

Fig. 2 Pitfall trapping plots, Haleakala volcano, Maui, 2002–2003, with populations of the unicolonial Argentine ant, *Linepithema humile*. Lower ant population surrounds Pu'u O Ili; upper ant population includes Kalahaku

Table 2 Frequencies of macropterous and brachypterous specimens of *Trechus obtusus* for three sites spanning six years on Haleakala volcano, Maui, and gene frequencies for macropterous (p) and brachypterous (Q) alleles, assuming single-locus, brachyptery-dominant basis of wing dimorphism (Lindroth 1946; Aukema 1990, Aukema et al. 1996)

Site and year	freq. macropters	freq. brachypters	p	Q
Service area				
1999	4	0	1.0	0
2001	6	0	1.0	0
2002	49	24	0.82	0.18
Pu'u O Ili				
2002	85	4	0.977	0.022
2003	247	2	0.996	0.004
Kalakahu				
2002	4	0	1.0	0
2003	1	0	1.0	0
Polipoli				
2001	96	0	1.0	0
2003	48	1	0.99	0.01
2005	124	22	0.92	0.08

year (2002, 2003), plus their interaction. The response variables (beetle catches) were ln+1 transformed to make variances equal for all groups (according to Levene's test) and to decrease heteroscedasticity of the residuals.

Biological specimens

All carabid beetle specimens are deposited in institutional collections; JKL in the Cornell University Insect Collection (CUIC), and PK in the Bernice P. Bishop Museum. In addition, specimens collected by R. Takumi Kaholoa'a are deposited in the Haleakala National Park Insect Collection and the Bishop Museum. *Trechus obtusus* diagnosis and identification was presented in Liebherr and Takumi (2002). Native Carabidae reported herein from the study area consist of species of *Blackburnia* Sharp (Liebherr and Zimmerman 2000), and *Mecyclothorax* Sharp (Britton 1948; Liebherr 2005).

Trechus obtusus exhibits wing dimorphism (den Boer 1968; Desender et al. 1980a, b), with populations often consisting of a mixture of individuals possessing macropterous metathoracic flight wings, and brachypterous metathoracic alae that vary in length, the apex extending to above the apical margin of the second to third abdominal tergite. As far as is known, carabid beetle wing dimorphism is controlled by a single allele, with the macropterous morph homozygously recessive and brachypterous individuals either heterozygous or homozygous for the dominant brachypterous allele (Lindroth 1946; Aukema 1990; Aukema et al. 1996). Beetles from all samples were examined for flight wing configuration by gently lifting one elytron, or by viewing the folded macropterous wings held against the translucent elytra. *Trechus obtusus* overwinters as both larvae and adults in Europe (Lindroth 1992). Preliminary data on this species' life history on Haleakala were gathered by noting the presence of teneral specimens among the various samples gathered by pitfall trapping and leaf litter sifting. The presence of tenerals was interepreted as evidence that adult beetles had eclosed recently, and that mature larvae had been present approximately one month prior to collection of the teneral adults.

Results

Trechus and Mecyclothorax

The initial 2001 visit to the gulch site resulted in the discovery of 77 *Trechus obtusus* within the four m^2 samples of forest floor leaf litter, complemented by 20 specimens of three different native *Mecyclothorax* species (Table 1). Given that this initial visit was aimed at gathering taxonomic specimens, the four samples were pooled, leaving only the totals available for analysis. In 2003, five samples were maintained as distinct entities, though there was little difference among them. In all five samples, *T. obtusus* predominated, with only a single specimen of one of two native species in three of the five samples. The third visit in 2005 resulted in total numbers of beetles in each sample much like those of 2003, though native *Mecyclothorax* constituted 13% of the beetles found versus 6% of beetles found in 2003. The percentages of native beetles in 2003 and 2005 were both lower than the 21% native beetles found in 2001. That the sampling strategy did not underestimate native beetles is supported by the results of the 2005 "taxonomist's" lot, which focused on situations within which experience suggested that native beetles would be found; i.e. moister microhabitats around the bases of dense *Dryopteris* fern clumps. In these situations, *T. obtusus* dominated, with native beetles accounting for only 6% of the specimens.

A χ^2-test, excluding the 2005 "taxonomist's" lot from the analysis, results in a significant χ^2-value of 16.97 (d.f. = 8, $0.05>P>0.025$), supporting the contention that relatively fewer native beetles were present on the gulch forest floor in 2003 and 2005. Contributions to the significance of the test are concentrated in cells represented by *M. cordithorax*.

During each yearly sampling effort, from 1 to 2% of the gulch bottom surface was raked and sifted for beetles. That low percentage coupled with the two-year intervals between the samples is assumed to have established independence of the samples. If that were not true, we might expect a diminution from sample to sample of the more rarely encountered species of *Mecyclothorax*. Aside from the falloff in numbers of the most common *M. cordithorax* from 2001 versus 2003 and 2005, this was not seen. Conversely, the 2005 samples contained the highest numbers of *M. consobrinus* Liebherr and *M. irregularis* Britton, and the only collection of *M. aeneipennis* Liebherr in the gulch. Given the greater sampling effort in 2005, care must be taken in comparing total numbers and diversity. That effect can be controlled by restricting diversity comparisons to the first five chronologically taken 2005 samples (lots 01–05). These lots contain more *M. consobrinus* and *M. irregularis* than observed in the five 2003 samples, strongly suggesting that prior sampling had no demonstrable influence on subsequent resident beetle diversity.

Wing dimorphism within *Trechus obtusus*

The initial samples of *Trechus obtusus* collected in 1999 and 2001 and reported in Liebherr and Takumi (2002) comprised exclusively macropterous individuals (Table 2). In 2002, the first brachypterous *Trechus* individuals were collected on Maui, both at the Service Area and near Pu'u O Ili, though the elevationally higher Pu'u O Ili site exhibited a lower percentage of brachypterous individuals for both 2002 and 2003. The relatively distant Polipoli Springs State Recreation Area population (Fig. 1) transformed from exclusively macropterous in 2001, to nearly exclusively macropterous in 2003 – with only a single brachypterous individual among 49 specimens – to dimorphic, with 15% of the individuals brachypterous in 2005. At the Service Area, the frequency of the brachypterous allele has increased from a putative 0% to 18% between 1999 and 2002, whereas the Polipoli population has undergone biennial increases in the brachypterous allele from 0 to 1%, to 8% (Table 2).

Two final samples, though consisting of only one beetle each, were collected in November, 2005. The specimen collected on Hanakauhi Peak is macropterous, however the singleton collected in Koolau Gap is brachypterous. Though only a single specimen, this latter sample expands the known distribution of the brachypterous morph on the mountain by 4 km toward the east from the putative point of origin of the invasion nearest the Service Area.

Trechus, Mecyclothorax and Ants

Only 2 *Mecyclothorax* individuals (1 in 2002 and 1 in 2003) and 4 *Trechus* individuals (3 in 2002 and 1 in 2003) were captured at the higher elevation Kalahaku site using the comparative abundance plot design. While it is notable that both *Mecyclothorax* beetles were captured outside the ant range, and 3 of 4 *Trechus* beetles were captured inside the ant range, numbers at the Kalahaku site were too low for statistical analysis.

Results of the GLM for *Mecyclothorax* species collected at the lower Pu'u O Ili site (including *M. cordithorax*, *M. montivagus* and *M. apicalis*) indicate that the factor *ant presence* had a significant effect on natural log -transformed *Mecyclothorax* abundance ($F_{1,31}$=7.29, P=0.012, Table 3); there were significantly fewer *Mecyclothorax* within the ant range as compared to outside the ant range. The factor *year* had no significant effect on *Mecyclothorax* abundance ($F_{1,31}$=0.36, P=0.553), and the interaction between *ant presence* and *year* was also not significant ($F_{1,31}$ = 0.08, P = 0.784). Numbers of *M. apicalis* and *M. montivagus* captured were too low (3 and 1 individuals, respectively) to analyze separately, however both of these species were only collected outside the ant range. *M. cordithorax*, which accounted for almost all of the captures in both years, was the only *Mecyclothorax* species to be collected inside the ant range. All captures of this species inside the ant range occurred in one gulch.

Results of the GLM for *Trechus* indicate that none of the factors in the model had significant influences on natural log -transformed *Trechus* abundance ($F_{1,31}$ = 0.00, P = 0.985 for *ant presence*; $F_{1,31}$ = 0.04, P = 0.834 for *year*; $F_{1,31}$ = 0.01, P = 0.910 for *ant presence × year*). There was therefore no significant impact of Argentine ants on *Trechus* numbers in either year. *Trechus* captures were highest within the ant range, in the same gulch where *M. cordithorax* was found.

Table 3 Mean (± SE) beetle captures per plot in comparative abundance pitfall sampling at Pu'u O Ili site. *Mecyclothorax* species included *M. cordithorax*, *M. montivagus* and *M. apicalis*

Year	Ant presence	*Trechus obtusus*	*Mecyclothorax* spp.
2002	ants present	5.63 (±3.38)	0 (±0)
	ants absent	4.25 (±2.74)	3.50 (±2.67)
2003	ants present	8.88 (±6.84)	0.75 (±2.12)
	ants absent	3.13 (±1.25)	2.75 (±1.37)

GLM test indicated that ant presence acting on natural log -transformed *Mecyclothorax* abundance ($F_{1,31}$=7.29, P=0.012) is only significant factor (see text)

Discussion

Natives and aliens in an exotic ecosystem

The afforested Kula Forest Reserve near Polipoli exhibits the greatest level of sympatry between alien *Trechus obtusus* and multiple native species of *Mecyclothorax* beetles. This forest has been extensively and dramatically modified from its native condition. The area presently consists of plantations of Monterey Pine, Arizona Cypress (*Cupressus arizonica*), Redwood (*Sequoia semperivirens*) and Sugi (*Cryptomeria japonica*), among other gymnosperms (Little and Skolmen 1989). These plantations were established to combat the extensive erosion of the drier slopes of Haleakala volcano brought on by cattle grazing (Giffard 1918; Cuddihy and Stone 1990). During the early 20th Century on Maui, "The almost total destruction of the undergrowth has allowed the soil to bake and harden thus causing the rainfall to run off rapidly... The district of Kula is also a striking example and, in order to save the little remaining forest, the cattle must be absolutely excluded (Griffith 1902)." The gulch site within which relative abundances of *Trechus* and *Mecyclothorax* were compared contains native elements such as *Dryopteris* ferns and overstory *Ilex* and *Myrsine* trees. However four native *Mecyclothorax* species – *M. aeneipennis*, *M. consobrinus*, *M. cordithorax*, *M. ovipennis* Sharp – are also found under bark of non-native Monterey Pine and Arizona Cypress in company with other native species of *Oodemas* weevils (Curculionidae) and *Dromaeolus* eucnemid beetles (Eucnemidae) (Liebherr 2005). In the case of *M. consobrinus* and *M. aeneipennis*, these native species are known only from this restricted afforested area (Liebherr 2005). This finding belies the absolute restriction of native Hawaiian insects to native ecosystems due to some intrinsic competitive disadvantage.

We have only limited historical perspective on species abundances for native *Mecyclothorax* at Polipoli. Only W. M. Giffard visited the site in 1913 prior to modern-day sampling (Liebherr 2005). Giffard's collections included seven specimens: 1, the unique holotype of *Blackburnia polipoli* Liebherr (Liebherr and Zimmerman 2000); 2, the unique holotype of

M. giffardi Liebherr; 3, three specimens of *M. consobrinus*; and 4, two specimens of *M. cordithorax* (Liebherr 2005). That the first two species have not been recovered recently makes them unknown quantities in the biodiversity calculus of the Kula Forest Reserve. *Mecyclothorax consobrinus* has been extensively re-collected at a variety of elevations on the Polipoli area, suggesting that its ecological preferenda may be broader than those of the other two species. *Mecyclothorax cordithorax* was very rarely collected prior to the 1960s, but is now the most commonly encountered *Mecyclothorax* in mesic to dryland habitats across leeward Haleakala and in the caldera crater (Liebherr 2005). For this species, a significant, dramatic increase in relative abundance appears undeniable.

Limited arthropod surveys in non-native dominated forest habitats have been performed in Hawaii. Agricultural areas were surveyed for native species in lowland Kauai, with 24 of the 283 species found determined to be endemic to Hawaii (Asquith and Messing 1993). Nine of these endemic species were associated with stream and riparian habitats that traversed the agriculturally modified fields. Given that agricultural production brings continuous disturbance to such situations, the endemic 9% of the fauna would appear to represent a robust element in such a local lowland fauna. The native fauna of Polipoli would appear even more likely to persist assuming no greater degradation of this novel habitat.

Based on available data for *T. obtusus* and *Mecyclothorax* species at the Polipoli gulch site, a temporal trend toward relative reduction in the native component of the forest-floor carabid community is indicated. The significance of the reduction rides on the decrease in representation of *M. cordithorax* in samples from 2003 and 2005 versus 2001 when *T. obtusus* was first discovered (Table 1). As *Mecyclothorax cordithorax* is found in shrubland habitats across Haleakala as well as within the mesic, closed canopy afforested areas near Polipoli (Liebherr 2005), the relative diminution in abundance of this species at the gulch site does not accord with its broader habitat preference. Drought as a factor negatively impacting beetle populations also does not accord with available climatological data. The Polipoli Springs weather station (NOAA 2005) recorded 591 mm of rainfall during the 12-month period prior to May 2001 when 4 m^2 samples resulted in 97 beetles. The much smaller 2003 samples comprising 49 beetles were preceded by a 12-month period during which 847 mm of rain fell. Alternately considering that population levels might be dependently determined over longer periods we could expand the considered rainfall window to 24 months prior to sampling. However this also does not remove the disparity; as 985 mm of rain fell prior to the 2001 samples versus 2070 mm prior to the smaller 2003 samples.

If the relative abundance of native versus introduced carabid beetles is based on their biotic interactions, both the predatory adult and larval stages may be likely participants in the biotic interactions. Landry (1994) hypothesized that observed differences in reproductive period and adult emergence governed the differential composition of species assemblages comprising closely related marsh-inhabiting carabids in Alberta. Species with smaller-bodied adults coexisted with larger bodied taxa, with larvae of the smaller bodied taxa eclosing earlier, and thereby being larger at the point of their generation when they became syntopic with larvae of later developing, ultimately larger-bodied taxa. *Trechus obtusus* adults are smaller than adults of the gulch-site *Mecyclothorax* species, suggesting their larvae could be suitable prey items for *Mecyclothorax* larvae and adults. The degree of synchrony among the various life stages of *Trechus* and *Mecyclothorax* must be determined, however, to document possible predatory interactions. To date, *Trechus* samples from Polipoli, collected in May, contained up to 20% teneral individuals (Table 1); a result consistent with larval overwintering, early spring pupation, and subsequent emergence of adults in early May. However, January to March 2002 samples of *T. obtusus* from the Service Area comprised exclusively sclerotized adults, indicating that adults are overwintering at this site on Haleakala. Thus the life cycle of *T. obtusus* on Haleakala appears to parallel that in Sweden, where both adult and larval stages overwinter (Lindroth 1992). All May-collected Polipoli specimens of *M. cordithorax* (36 exx.), *M. consobrinus* (25 exx.), and *M. irregularis* (6 exx.) (Table 1, Liebherr 2005) are fully sclerotized, indicating adult overwintering. Two of 13 May specimens of *M. aeneipennis* are teneral, however, suggesting that this species may overwinter in part in the larval stage, similar to *T. obtusus*. In addition to any differences or similarities in life cycle timing and their affects on biotic interactions among *Trechus* and *Mecyclothorax* individuals, we need to learn whether the large numerical dominance of *Trechus* oberservable at Polipoli is based on an intrinsically higher rate of increase that might lead to an advantage via mutual predation or direct competition. Such comparative data are available for another carabid beetle that has invaded Hawaii – *Metacolpodes buchanani* (Hope) (Liebherr and Zimmerman 2000) – with this newly arrived species exhibiting laboratory fecundity levels (Paarmann and Bolte 1990) more than

10× that recorded for contribal, native Hawaiian *Blackburnia* species (Liebherr 2000).

Adaptation of Trechus

The dramatic increase in the frequency of brachyptery among *Trechus obtusus* beetles at the Service Area and Polipoli sites (Table 2) is not entirely unprecedented. Den Boer et al. (1968) found that 95% of *T. obtusus* individuals were macropterous in 1964 samples from a Flevoland polder reclaimed from the sea seven years earlier. These results contrasted with a second, long natural site near Drenthe where only 4% of the individuals were macropterous. As done in this study, assuming the only demonstrated genetic mechanism for carabid wing dimorphism, the change in frequency of the recessive macropterous allele at the Flevoland polder would have been about 0.025 in seven years, assuming the initial propagule colonized the reclaimed site by flight. Desender et al. (1980a, b) surveyed wing dimorphism in *T. obtusus* across an ecological transect ranging from unstable dunes to coastal forest in the Westhoek Nature Preserve along coastal Belgium, finding that percentage macroptery ranged from 100% in the most unstable habitats to 0.5% at the forest sites. Thus this species can adapt quickly to extremely localized situations, with the flightless forms predominating in any stable habitat (Southwood 1977).

The most dramatic increases in frequency of the brachypterous morphs of *T. obtusus* have occurred at both the Service Area and Polipoli (Table 2). Based on the initial 1999 discovery at the Service Area and contemporaneously documented absence of this species from Polipoli (Liebherr & Takumi 2002), the Service Area appears – among those localities known to house *T. obtusus* in Maui – as closest to the initial area of introduction. Based on presence of this species at Puu O Ili and Kalahaku (Fig. 1), this species is currently continuously distributed across the leeward Halakala face, and thus it would appear to have rapidly expanded its range to Polipoli. Such rapid expansion may have occurred via active winged flight, or passively via human transport. As this species colonized Maui via human transport, this possibility can never be discounted. The recent discovery of a single brachypterous specimen of *T. obtusus* in Koolau Gap, more than 4 km east of the Service Area, points to inadequacies in our knowledge of the distribution of *T. obtusus* across the mountain. If we cannot accurately circumscribe the distribution of *T. obtusus*, we cannot make conclusions about how fast its distribution is spreading. Relative to Polipoli, we can only report its absence in 1999 and its presence in 2001.

Based on samples from 1999 and 2001, Liebherr and Takumi (2002) concluded that the initial Maui *T. obtusus* propagule was 100% macropterous (Table 2). They thereby proposed both a source for the introduction – Oregon or the San Francisco Bay Area – and an initial propagule size; less than 25 individuals, based on matching gene frequencies from previously reported mainland populations (Kavanaugh and Erwin 1985). Viewing the small 1999 and 2001 Service Area samples with hindsight provided by the rapid increase in frequency of the brachypterous allele in the 2002 sample, we must reject both conclusions as untenable. Given patterns of commercial shipping to Hawaii (Bess and Marcus 1998), a point of origin on the United States Pacific Coast is most reasonable, but any more specific conclusion regarding geographic source based on presently available data is unsupportable.

The consequences of the evolution of brachyptery in *Trechus obtusus* may be several. As this species colonizes suitable habitats on Maui and the frequency of brachyptery increases, the incidence of winged dispersal should decrease. Moreover, gene frequencies in established populations will be little influenced by the contributions of subsequent dispersants. To the degree that localized conditions differentially influence survival of various phenotypes, local populations may diversify with regard to phenology. From a conservation standpoint, these events would reduce the possibility of natural winged dispersal from Maui to the other Hawaiian Islands. However, the possibility of transport by human means will remain, and as this was the means of entry into Maui, only careful inspection of the types of freight inhabited by *Trechus* – especially nursery stock (Kavanaugh and Erwin 1985) – will prevent eventual spread of this species to the other islands. If brachypterous *Trechus* individuals exhibit greater fecundity than the colonizing macropters (Denno et al. 1989), increased representation of *Trechus* in habitats originally housing only native *Mecyclothorax* species may follow. Polipoli will be one of the best sites to observe any further changes in relative abundances of *Trechus* versus a variety of native *Mecyclothorax*. The consequence of macropterous individuals, or at least alleles, remaining in *Trechus* populations will ensure that this species will be able to colonize novel habitats through persistent production of flight-capable individuals. On Maui therefore, *T. obtusus* will quickly become more widespread than any of the monomorphically brachypterous *Mecyclothorax* species (Britton 1948; Liebherr 2005) – none of whom are found on both Haleakala and West Maui (Liebherr unpubl. data) – resulting in both larger and more stable *Trechus* populations, as well as a more dynamic metapopulational structure.

Mecyclothorax, Trechus and ants

Mecyclothorax individuals, comprising mostly *M. cordithorax*, occurred in significantly fewer numbers within ant-infested shrubland as compared to adjacent ant-free areas at the Pu'u O Ili site on Haleakala's northwest slope. Though too few to analyze separately, *M. montivagus* and *M. apicalis* individuals were only captured outside Argentine ant population boundaries. At the higher elevation Kalahaku site, one individual each of *Mecyclothorax micans* (Blackburn) and *Mecyclothorax rusticus* Sharp were also only collected outside ant population boundaries. These results, together with earlier findings (Loope et al. 1988; Cole et al. 1992), suggest that most *Mecyclothorax* species in Haleakala National Park are strongly impacted by Argentine ant invasion. *M. cordithorax* also occurs less abundantly within ant-infested areas, but this species can apparently persist, at least for some time, in the presence of ants in localized microhabitat pockets such as wet gulches.

Conversely, the presence of *T. obtusus* both inside and outside the infestation areas of the Argentine ant suggests that this species can cope with large numbers of invasive ants. This finding is in line with results from three different continental communities in California, USA, where carabid beetle abundance and diversity was either significantly greater (Human and Gordon 1997; Bolger et al. 2000) or not significantly different (Holway 1998) in areas invaded by Argentine ant. In San Diego Co., CA, carabid beetles most commonly found in significantly greater numbers in the company of Argentine ants included the adventively introduced, North African species *Laemostenus complanatus* (Dejean) (Casale 1988), and two native Californian species; an *Hypherpes isabellae*-complex species, and *Calathus ruficollis* Dejean (A. V. Suarez and K. W. Will, pers. comm.). Two of these three – *L. complanatus* (Zimmerman 1972) and *C. ruficollis* (Liebherr and Zimmerman 2000) – have become established in Hawaii. All mainland carabid species found in greater numbers inside ant-infested areas have evolved in regions housing native ants. In Hawaii, the persistence of *Trechus* in the company of ants, at the same time that native *Mecyclothorax* nearly disappear, could be attributed to four possible factors: 1, competitive superiority of individuals in the various life stages of *Trechus* relative to *Mecyclothorax* when directly interacting with predaceous ants; 2, temporal avoidance of ants by *Trechus* individuals; 3, spatial avoidance of ants by *Trechus* individuals; 4, greater reproduction or immigration of *T. obtusus* relative to *Mecyclothorax* as a means of sustaining populations in the face of ant predation.

The first option is unlikely because *Trechus* larvae (Wallin 1989) do not exhibit the specialized decoy or appeasement behavior observed in myrmecophiles (DiGiulio and Vigna Taglianti 2001) or ant predators (Dinter et al. 2002). Moreover, carabid larvae pupate in secretive situations by modifying small voids in soil or humus, but otherwise do not exhibit behavior that would defend against ant predation. The adults possess pygidial defensive glands that can be used to deter predators, yet adults of non-myrmecophilous carabid taxa occur less abundantly close to actively foraging colonies of dominant ant species (e.g. Hawes et al. 2002).

The plasticity apparent in the life cycle of *T. obtusus* may afford some temporal isolation from Argentine ant. The presence of teneral adults in mid-May suggests some of the individuals overwinter in the larval stage, as typical in parts of the European range (Lindroth 1992). However, the numerous sclerotized adults pitfall-trapped in January to March indicate that adult individuals also overwinter, as observed in Sweden (Lindroth 1985). Those larvae present late in the summer season, and persisting through winter would be afoot when ant foraging would be at its lowest level due to ambient conditions. Moreover, pupation in early spring would occur before ant populations could increase during the summer foraging season (Krushelnycky et al. 2004). The temporal distribution of *Trechus* larvae must be determined to evaluate this hypothesis.

Microhabitat preference could also afford spatial isolation of *Trechus* from Argentine ant foraging. Though data are not available on the distribution of *T. obtusus* larvae throughout the soil layers, larvae of *T. secalis* (Paykull) were found to be completely subterranean in arable fields in Sweden (Wallin 1989). This microhabitat preference contrasted with syntopic larvae of *Bembidion lampros* (Herbst), which also were active on the soil surface. A comparative study of the temporal and spatial activity patterns of *T. obtusus* versus *Mecyclothorax* larvae could provide a valuable predictor of larval-ant interactions.

Fourthly, it is possible that *Trechus* is more successful in the presence of ants relative to *Mecyclothorax* because of higher reproductive output or the ability of winged individuals to enter ant-infested areas. *Trechus* greatly outnumbered *Mecyclothorax* at the Polipoli gulch site, but so far there has been only a slight (and probably statistically non-significant) numerical advantage over *Mecyclothorax* in non-ant-invaded plots at the Pu'u O Ili site. Small differences in reproduction, however, may lead to larger population asymmetries over time, and may be amplified in importance in the context of competition with or predation by ants. *Trechus* abundances were highest in the same gulch that served as the only detected refuge for *M. cordithorax* within ant-infested habitat. The combination of invasive ants and large numbers of competing invasive carabids, therefore, could eventually make this type of microhabitat uninhabitable for *M. cordithorax*.

The Hawaiian *Mecyclothorax* radiation has its roots in Australia, where *Mecyclothorax* beetles occur in riparian and forest-floor habitats also housing ants. Britton (1948) proposed the Australian *Mecyclothorax punctipennis* (MacLeay) as the closest patristic relative to the Hawaiian *Mecyclothorax* radiation. The Hawaiian species most similar to *M. punctipennis* include *M. cordithorax* and *M. montivagus*, both from East Maui, making Maui the logical point of initial Hawaiian colonization by *Mecyclothorax* some time during the past million years or so. In contrast to the Hawaiian *Mecyclothorax*, *M. punctipennis* is a flight-capable species with a broad geographic range that spans Australia (Moore 1984). From this disparity come two alternate hypotheses. First, Hawaiian *Mecyclothorax* may lack some attributes present in Australian *Mecyclothorax* that allow the Australians to coexist with ants. Alternatively, Australian *Mecyclothorax* and Hawaiian *Mecyclothorax* may be equally susceptible to ant predation, but the volant Australian taxa may repeatedly colonize localized situations not dominated by ants, thereby escaping, for a time, ant predation. Given this second alternative, we might also hypothesize that the Hawaiian *Mecyclothorax* ancestral lineages underwent ecological release due to the absence of ants, allowing evolutionary movement into dry-land habitat types that remain dominated by ants in Australia. Hawaiian *Mecyclothorax* also underwent conversion to brachyptery, limiting their ability to colonize disjunct habitats. Only now are these highly endemic dry-land taxa facing the resumption of ant predation and competition. This latter interpretation is consistent with the much greater species richness and geographical provincialism of the Hawaiian *Mecyclothorax* (Britton 1948; Liebherr 2005) compared to the much older Australian fauna that comprises only 18 described species, nearly half geographically widespread (Moore et al.1987; Baehr 2003).

Acknowledgements This research was supported in part by the National Science Foundation REVSYS award DEB-0315504, the Hawaii Audubon Society, Pacific Rim Research Program, the Margaret C. Walker Fund, the National Park Service Inventory and Monitoring Program, Haleakala National Park and a National Science Foundation Graduate Fellowship to P.D.K. We thank Raina Takumi Kaholoa'a, Haleakala National Park, for sampling carabid beetles, and Rosemary Gillespie for providing guidance to P.D.K. Mandy Anhalt, Caroline Berman, Jason Long, Marshall Loope, Andrew Marks and Kimberly Tice assisted with sorting pitfall samples. Betsy Harrison Gagné, Natural Areas Reserve System, State of Hawaii Department of Land and Natural Resources provided assistance and encouragement for surveying the insect fauna of Hawaii. We thank Forest Starr for a copy of E. M. Griffith's 1902 report on the status of Hawaiian Forests, and Andy Suarez and Kip Will for sharing their San Diego Co. carabid beetle data. We also thank Dan A. Polhemus and Curtis P. Ewing for their collaborations in field surveys. The Pacific Cooperative Studies Unit, Department of Botany, University of Hawaii at Manoa, provided administrative assistance.

References

Asquith A and Messing RH (1993) Contemporary Hawaiian insect fauna of a lowland agricultural area on Kauai: implications for local and island-wide fruit fly eradication programs. Pac Sci 47:1–16

Aukema B (1990) Wing-length determination in two wing-dimorphic *Calathus* species (Coleoptera: Carabidae). Hereditas 113:189–202

Aukema B, Spee AJ and van Dijk TS (1996) Wing dimorphism and development in *Pterostichus melanarius* (Coleoptera: Carabidae). Entomol Ber Amsterdam 56:93–100

Baehr M (2003) Psydrine ground beetles (Coleoptera: Carabidae: Psydrinae), excluding Amblytelini, of eastern Queensland rainforests. Mem Queensland Mus 49:65–109

Beardsley JW, Jr (1979) New immigrant insects in Hawaii: 1962–1976. Proc Hawaii Entomol Soc 23:35–44

Bess D and Marcus H (1998) Transportation. In: Juvik SP and Juvik JO (eds) Atlas of Hawaii. University of Hawaii Press, Honolulu, pp 268–272

Bolger DT, Suarez AV, Crooks KR, Morrison SA and Case TJ (2000) Arthropods in urban habitat fragments in southern California: area, age and edge effects. Ecol Appl 10: 1230–1248

Britton EB (1948) A revision of the Hawaiian species of *Mecyclothorax* (Coleoptera: Carabidae). Occ Pap B P Bishop Mus 19:107–166

Casale A (1988) Revisione deglie Sphodrina (Coleoptera, Carabidae, Sphodrini). Mus Reg Sci Natur Torino Monogr 5:1024

Cole FR, Medeiros AC, Loope LL and Zuhlke WM (1992) Effects of the Argentine ant on arthropod fauna of Hawaiian high-elevation shrubland. Ecology 73:1313–1322

Cuddihy LW and Stone CP (1990) Alteration of native Hawaiian vegetation: effects of humans, their activities and introductions. University of Hawaii Press, Honolulu

den Boer PJ (1968) Zoölogisch onderzoek op het Biologisch Station, Wijster 1959–1967. Misc Pap Landbouw Wageningen The Netherlands 2:161–181

Denno RF, Olmstead KL and McCloud ES (1989) Reproductive cost of flight capability: a comparison of life history traits in wing-dimorphic planthoppers. Ecol Entomol 14:31–44

Desender K, Maelfait J-P and Duerinck R (1980a) Ecological data on *Trechus obtusus* De Geer (Coleoptera, Carabidae) collected by pitfall trapping in coastal dunes (Belgium). Biol Jaarb Dodonaea 48:90–101

Desender K, Maelfait J-P, Vanhercke L, and Deurinck R (1980b) Investigations on Coleoptera communities in different habitats I. the carabid fauna of the dune nature preserve "De Westhoek" (De Panne, Belgium). Biol Jaarb Dodonaea 48:102–110

Di Giulio A and Vigna Taglianti A (2001) Biological observations on *Pachyteles* larvae (Coleoptera Carabidae Paussinae). Trop Zool 14:157–173

Dinter K, Paarmann W, Peschke K and Arndt E (2002) Ecological, behavioural and chemical adaptations to ant predation in species of *Thermophilum* and *Graphipterus* (Coleoptera: Carabidae) in the Sahara desert. J Arid Environ 50:267–286

Dobson AP, Rodriguez JP, Roberts WM and Wilcove DS (1997) Geographic distribution of endangered species in the United States. Science 275:550–553

Eldredge LG and Evenhuis NL (2003) Hawaii's biodiversity: a detailed assessment of the numbers of species in the Hawaiian Islands. Bishop Mus Occ Pap No 76:28

Englund RA (1999) The impacts of introduced poeciliid fish and Odonata on the endemic *Megalagrion* (Odonata) damselflies of Oahu Island, Hawaii. J Ins Conserv 3:225–243

Gambino P, Medeiros AC and Loope LL (1987) Introduced vespids *Paravespula pensylvanica* prey on Maui's endemic arthropod fauna. J Trop Ecol 3:169–170

Gambino P (1992) Yellowjacket (*Vespula pensylvanica*) predation at Hawaii Volcanoes and Haleakala National Parks: identity of prey items. Proc Hawaii Entomol Soc 31:157–164

Giffard WM (1918) An appeal for action on forestry work. Hawaii Planters' Rec 18:539–543

Gillespie RG and Reimer N (1993) The effect of alien predatory ants (Hymenoptera: Formicidae) on Hawaiian endemic spiders (Araneae: Tetragnathidae). Pac Sci 47:21–33

Griffith EM (1902) General description of the Hawaiian forests, a report of U.S. Forester E. M. Griffith on Hawaiian forests presented at Yokohama, Japan, 8 pp. (transcribed). (ms. preserved in headquaters office, State of Hawaii Department of Land and Natural Resources, Division of Forestry and Wildlife, Honolulu)

Hadfield MG, Miller SE and Carwile AH (1993) The decimation of endemic Hawai'ian tree snails by alien predators. Am Zool 33:610–622

Hawes C, Stewart AJA and Evans HF (2002) The impact of wood ants (*Formica rufa*) on the distribution and abundance of ground beetles (Coleoptera: Carabidae) in a Scots pine plantation. Oecologia 131:612–619

Henneman ML and Memmott J (2001) Infiltration of a Hawaiian community by introduced biological control agents. Science 293:1314–1316

Holway DA (1998) Effect of Argentine ant invasions on ground-dwelling arthropods in northern California riparian woodlands. Oecologia 116:252–258

Human KG and Gordon DM (1997) Effects of Argentine ants on invertebrate biodiversity in northern California. Cons Biol 11:1242–1248

Kavanaugh DH and Erwin TL (1985) *Trechus obtusus* Erichson (Coleoptera: Carabidae), a European ground beetle, on the Pacific Coast of North America: its distribution, introduction, and spread. Pan-Pac Entomol 61:170–179

Krushelnycky PD, Gillespie RG, Loope LL and Liebherr JK (2005) Rediscovery and uncertain future of high-elevation Haleakala carabid beetles (Coleoptera). Pac Sci 59:399–410

Krushelnycky PD, Loope LL and Joe SM. (2004) Limiting spread of a unicolonial invasive insect and characterization of seasonal patterns of range expansion. Biol Invas 6:47–57

Lach L (2004) Floral nectar foraging by invasive ants and consequent effects on seed set and floral arthropod visitors. Ph.D. dissertation, Cornell University, Ithaca, NY

Landry J-F (1994) Resource partitioning in a guild of marsh-dwelling *Agonum* (Coleoptera: Carabidae) in central. Alberta Can Entomol 126:709–728

LaPolla J, Otte D and Spearman LA (2000) Assessment of the effects of ants on Hawaiian crickets. J Orthopt Res 9:139–148

Liebherr JK (2000) The unity of characters: ecological and morphological specialisation in larvae of Hawaiian platynine Carabidae (Coleoptera). Invert Taxon 14:931–940

Liebherr JK (2004) Patterns of endangerment or pathways to enlightenment? Reconstructing the Fauna Hawaiiensis. Syst Biodiv 2:175–189

Liebherr JK (2005) New species of *Mecyclothorax* (Coleoptera: Carabidae, Psydrini) from Polipoli, Maui define an area of endemism on Haleakala volcano, Hawaii. J New York Entomol Soc 113:97–128

Liebherr JK and Takumi R (2002) Introduction and distributional expansion of *Trechus obtusus* (Coleoptera: Carabidae) in Maui, Hawaii. Pac Sci 56:365–375

Liebherr JK and Zimmerman EC (2000) Hawaiian Carabidae (Coleoptera), Part 1: Introduction and Tribe Platynini. Ins Hawaii 16:494

Lindroth CH (1946) Inheritance of wing dimorphism in *Pterostichus anthracinus* Ill. Heriditas 32:37–40

Lindroth CH (1985) The Carabidae (Coleoptera) of Fennoscandia and Denmark. Fauna Entomol. Scand. 15(part 1):1–225

Lindroth CH (1992) Ground beetles (Carabidae) of Fennoscandia, a zoogeographic study, part i, specific knowledge regarding the species. Smithsonian Institution Libraries, Washington DC

Little EL, Jr and Skolmen RG (1989) Common forest trees of Hawaii (native and introduced). US DA For Serv Agr Handb 679:321

Loope LL, Medeiros AC and Cole FR (1988) Effects of the Argentine ant on the endemic biota of subalpine shrubland, Haleakala National Park, Hawaii. In: Thomas LK (eds) Management of exotic species in natural communities, vol. 5. The US National Park Service and The George Wright Society, Fort Collins Colorado, pp 52–62

Moore BP (1984) Taxonomic notes on some Australian *Mecyclothorax* Sharp (Coleoptera: Carabidae: Psydrinae) and descriptions of new species. J Aust Entomol Soc 23:161–166

Moore BP, Weir TA and Pyke JE (1987) Coleoptera: archostemata, myxophaga and adephaga: rhysodidae and Carabidae. In: DW Walton (eds), Zoological catalogue of Australia. Australian Government Printing Service, Canberra, pp 17–366

N.O.A.A. (2005) Annual climatological summary, station 518070/99999, Polipoli Springs 267.2. National Climatic Data Center, Asheville, NC (http://nndc.noaa.gov/?http://ols/ncdc.noaa.gov/)

Paarmann W and Bolte H (1990) Studies on the biology of *Colpodes buchanani* (Hope) (Coleoptera, Carabidae), collected in Sulawesi rain forest (Indonesia) by applications of knock-down insecticides with a canopy fogger. In: Knight WJ and Holloway JD (eds) Insects and the rain forest of south east Asia (Wallacea). The Royal Entomological Society, London, pp 191–195

Polhemus D and Asquith A (1996) Hawaiian Damselflies, A Field Identification Guide. Bishop Musuem Press, Honolulu

Pratt LW and Gon SM, III (1998) Terrestrial ecosystems. In: Juvik SP and Juvik JO (eds) Atlas of Hawaii. University of Hawaii Press, Honolulu, pp 121–129

Reimer NJ (1994) Distribution and impact of alien ants in vulnerable Hawaiian ecosystems. In: Williams DF (eds) Exotic ants: biology, impact, and control of introduced species. Westview Press, Boulder, pp 11–22

Rutledge DT, Lepczyk CA, Jialong X and Liu J (2001) Spatio-temporal dynamics of endangered species hotspots in the United States. Conserv Biol 15:475–487

Ryan B, Joiner B and Cryer J (2005) Minitab handbook, 5th edn. Thomson Brooks/Cole, Belmont CA, 505 pp

Sakai AK, Wagner WL and Mehrhoff LA (2002) Patterns of endangerment in the Hawaiian flora. Syst Biol 51:276–302

Southwood TRE (1977) Habitat, the templet for ecological strategies?. J Anim Ecol 46:337–365

Wallin H (1989) Habitat selection, reproduction and survival of two small carabid species on arable land: a comparison between *Trechus secalis* and *Bembidion lampros*. Hol Ecol 12:193–200

Zimmerman EC (1972) On some foreign Carabidae established in Hawaii (Coleoptera). Proc Hawaii Entomol Soc 21: 291–294

J Insect Conserv (2007) 11:75–83
DOI 10.1007/s10841-006-9020-2

BEETLE CONSERVATION

Water beetles associated with reservoirs on Table Mountain, Cape Town: implications for conservation

Clive R. Turner

Received: 5 November 2005 / Accepted: 1 March 2006 / Published online: 2 November 2006

Abstract Table Mountain maintained internationally high global conservation priority with respect to the aquatic Coleoptera. The study site displayed an overall 77% endemism with 65% of species provisionally of high IUCN international conservation status. Reservoirs were shown to have a dramatic negative impact on the endemic and endangered water beetles of Table Mountain. This comprised of an overall species loss of 73%, all endemics, and a 64% loss of high conservation status species in the Hydraenidae, Dryopidae and Elmidae. Further hydrological alteration in the region was not recommended and the necessary inclusion of water beetles in future EIA assessments stipulated. The necessity for bioinventory methodology in addition to standard netting methodology was conclusively illustrated. A Balanced Scorecard approach to conservation decisions based on species level data including appropriate bioinventory survey methodology, recent survey data, IUCN status, and levels of endemism was proposed. The differentiation of species communities with altitude produced a division between upland and lowland communities on Table Mountain. Distinct differences between the reservoir impacted and natural upland water beetle communities were demonstrated.

Keywords Africa · Aquatic Coleoptera · Conservation · Reservoir · Dam · Environmental impact assessment

C. R. Turner (✉)
15 Forresters Drive, Woolwell, Plymouth, Devon PL6 7QA, UK
e-mail: clive.turner@excite.com

Introduction

Initiated in the late nineteenth century, there are now five reservoirs in existence on the upland slopes of Table Mountain supplying water to the conurbation of Cape Town, South Africa (CCT 2001; Pauw and Johnson 2002). The environmental impact of these reservoirs had not been previously investigated with respect to the aquatic Coleoptera. It seems likely further regional reservoir construction will occur given the predicted 50% increase in South African demand for water between 1996 and 2030 (Bassen et al. 1997). This study initiates the essential process of understanding the impact of reservoir construction on the endemic and endangered fauna of the region. The water beetle fauna of southern Africa is composed of a mixture of pan Afrotropical generalists with widespread distributions and a significant proportion of regional endemics. The South African water beetle fauna currently comprises of 50% endemic species and 27% with a provisional IUCN international conservation status of endangered or higher (Turner in prep. a). These statistics clearly indicate the conservation importance of the regional fauna. Recent taxonomic advances and faunistic expansion (i.e. Ribera et al. 2002; Komarek 2004; Perkins 2004a, b, 2005a, b) illustrate the current paucity of valid taxonomic and ecological data on this fauna. Conservation of this fauna is in its infancy and although dialogue and interest exists it is hampered by the limited co-ordination of research effort symptomatic of restricted resources (Turner 2004, 2005, 2006). Within the regional fauna of the Cape Peninsula, and more specifically Table Mountain, there are a number of water beetles considered to be of extremely high conservation interest and a number of species are endemic to this limited area. Curiously this highly restricted distribution is not clearly displayed in the Dytiscidae (Omer-Cooper 1962, 1965; Turner in prep. a) probably due to their superior dispersal abilities. However the Haliplid *Algophilus lathridoides* Zimmermann 1924 is known only from the type locality of Simons Town, Cape Peninsula (Guignot 1959; Gueorguiev 1967). Perkins and Balfour-Browne (1994) specifically commented on the richness of the Table Mountain water beetle fauna where the Hydraenids *Prostethops nitens* (Péringuey 1892), *Mesoceration concessum* Perkins & Balfour-Browne 1994, *Pneuminion balfourbrownei* Perkins 2004 (Perkins 2004a) and *Pneuminion semisulcatum* Perkins 2004 (Perkins 2004a) are confined to the mountain itself, whilst *Prostethops megacephalus* (Boheman 1851) and *Pterostethops equestris* Perkins & Balfour-Browne 1994 are restricted to the mountain and immediate area. The additional more widely distributed species from the mountain, *Parasthetops aeneus* Perkins & Balfour-Browne 1994, *Parasthetops nigritus* Perkins & Balfour-Browne 1994, *Mesoceration languidum* Perkins & Balfour-Browne 1994, *Pterostethops impressus* Perkins & Balfour-Browne 1994, *Discozantaena genuvela* Perkins & Balfour-Browne 1994, justified the statement "a paradigm of unique biodiversity worthy of protection" (Perkins and Balfour-Browne 1994) before additional species were added to this list (*Coelometopon blinkwater* Perkins 2005 (Perkins 2005b), *Oomtelecopon setosum* Perkins 2005 (Perkins 2005b) and *Nucleotops interceps* Perkins 2004 (Perkins 2004b)). Perkins and Balfour-Browne (1994) quoted Branch (1989) who commented on the danger to threatened species caused by reservoir creation on the mountain. Given the reliance of these species, of which 4 are Table Mountain endemics, 2 Table Mountain proximal endemics, 8 Cape endemics, on habitat continuity an assessment of the impact of the reservoirs and other water management schemes would be essential to understanding the implications of reservoir development in the region. This study was intended as a preliminary investigation into the response of water beetles to reservoir construction on Table Mountain and was carried out as part of an extended scientific visit to South Africa by the author in 1997.

Since its inception in 1994 by Chutter, the South African Scoring System (SASS) for river health assessment has become established in South Africa and most recently amended as SASS5 methodology (Dickens and Graham 2002). Despite the use of ecological reference conditions (Dallas 2000) this system does not profess to fulfill bioinventory requirements but is often misconceived as adequate for the purpose (pers. obs.), however, as the sole method employed it is unsuitable for bioinventory surveys (Turner in prep b). SASS is intended for use as a user-friendly comparative benchmarking methodology for biomonitoring across South Africa (Day 2000) and indeed is the national standard and forms an integral part of the National River Health Programme (Uys et al. 1996). The net reliant SASS methodology was included in this survey and complemented with additional techniques. The impact of reservoir construction has been shown to create changes in physicality (i.e. Dadzie 1980; Beadle 1981; Davies & Day 1998) and faunistic composition (i.e. Dadzie 1980; Beadle 1981; Jackson et al. 1988; Gourène et al. 1999; Da Costa et al. 2000; Madikizela et al. 2001; Gratwicke et al. 2003; Madikizela and Dye 2003) resulting in its appearance as a factor in conservation assessments (i.e. Skelton 2001). The belated impact of reservoir construction on stream Coleopteran diversity above and below a reservoir in an area of international conservation priority remained unquantified prior to this study.

Methods

In February 1997, the Back Table of Table Mountain was visited to execute a preliminary assessment on the effect of reservoir presence on the aquatic Coleoptera in the vicinity of the Hely-Hutchinson Reservoir, the oldest reservoir on the mountain. The sample locations consisted of six sites (Table 1), and comprised one site above the reservoir (ahh), one in the reservoir (ihh) and one below the outlet of Woodhead Reservoir (bhh) because the outlet of the Hely–Hutchinson reservoir fed directly into Woodhead Reservoir below it. The outlet released water from the Woodhead reservoir into the natural channel of the Disa River valley thus continuing the historical flow of the river. In addition an adjacent, undammed, control site (vrg) was chosen for comparative purposes. A further two sites were selected where the outlet stream passed through the lowland indigenous forest. All sites were on the southern side of Table Mountain, with the Valley of the Red Gods being the only location not ultimately draining directly south. Given the variable nature of the terrain the sites were of an acceptably similar altitudinal range, with the upper mountain sites within the 720–780 m range and lower two sites adjacent and therefore of an identical altitude.

Sampling methodology was consistent across the three sites, based principally on the SASS methodology and supplemented with additional techniques

Table 1 Site data

Ref.	Locality name	Description	Latitude/Longitude
ahh	above Hely-Hutchinson	Disa River formed a small Fynbos surrounded stream of pools and riffles with Ericaceous and Fynbos decaying vegetative matter in the slow flowing pools. Substrate was a combination of rocks and gravel with areas of coarse sand. Altitude: c. 750 m asl.	18°24′37″ E 33°58′42″ S
ihh	in Hely-Hutchinson	The rocky margins of the reservoir were generally poorly vegetated, lacustrine, the substrate was essentially rocky with limited fine deposits and coarse sand. Altitude: c. 750 m asl.	18°24′33″ E 33°58′45″ S
bhh	below Hely-Hutchinson & Woodhead Reservoirs	A fast flowing, narrow channel, unvegetated environment with negligible vegetative deposits or finer sediments, rocky substrate forms the continuance of the Disa River. Altitude: c. 720 m asl.	18°24′06″ E 33°58′65″ S
vrg	Valley of the Red Gods	A narrow tributary of the Kasteelspoort flowing through the Valley of the Red Gods, a Fynbos surrounded stream of pools and riffles with Ericaceous and Fynbos decaying vegetative matter in the slow flowing pools. Substrate was a combination of rocks and gravel with areas of coarse sand. Altitude: c. 780 m asl.	18°23′47″ E 33°53′35″ S
dis	Disa River	The western tributary to the Hout Bay River. Indigenous forest surrounded lowland river with fast flowing, wide channel composed of an extremely coarse rocky substrate and little sedimentary deposits other than occasional areas of coarse sand, some aquatic vegetation. Altitude: c. 150 m asl.	18°23′47″ E 33°55′91″ S
ori	Original Disa Stream	Indigenous forest surrounded lowland river with fast flowing, wide channel composed of an extremely coarse rocky substrate and little sedimentary deposits other than occasional areas of coarse sand, some aquatic vegetation. The eastern tributary to the Hout Bay River. Altitude: c. 150 m asl.	18°23′50″ E 33°55′91″ S

Locality names and co-ordinates obtained from Slingsby (2004), MCSA (1993) and Pauw and Johnson (2002).

appropriate for bioinventory assessment. The limited size of the upland streams restricted the amount of quantitative sampling carried out. Hence the regime of two one minute samples with a standard square frame pond net (30 × 30 cm, 1 mm mesh), one minute in the riffles and margins with foot agitation where appropriate and the second minute in the pools and their margins with foot agitation where appropriate. All pond net samples were sorted in the field on a plastic sheet. Additional samples comprised rock turning, the Gyrinid skim (where Gyrinids visible), shore washing where appropriate, visual and hand searching of marginal habitats accompanied by stamping and splashing where appropriate. These methods are fully described and formalised in Turner (in prep b). Other than the standardised timed samples the remaining methods were continued until no new species were encountered. Material collected was preserved in screw top plastic tubes with the initial fixative, the locally available, 43% ethanol Cane Spirit and the material was later fully fixed with subsequent treatments of 70% ethanol. This method rendered material more relaxed and facilitated dissection (Turner in prep b). Whilst this study focussed on the aquatic Coleoptera the qualitative abundance of other taxa was noted to compliment the primary data.

The prevailing conditions were considered generally poor for aquatic invertebrates. Due to the summer seasonal reduction in precipitation there were reduced water levels with flows at their annual low. This was not considered an issue for the purposes of this study due to the comparative nature of the exercise, but for a comprehensive bioinventory it may be necessary to sample across seasons—an approach that may prove valuable in future studies. Seasonality tends to alter species distributions and to a lesser extent alters species composition but is not necessarily presented as a negative reaction to the dry season (pers. obs.).

Provisional international IUCN criteria were applied (Turner in prep. a) conforming with the international IUCN categorisation and criteria according to the revised version that included the new Critically Endangered category identified by Mace and Stuart (1994). The criteria contain varying stipulations some of which were inappropriate to most insects (Foster pers. comm.). The inappropriate criteria, usually related to ecological data based on census or predictions, have been excluded following Foster (pers. comm.). The provisional classification for each species was used to evaluate the conservation implications of water beetle community change with reservoir construction. Additionally the endemic status of the species encountered was considered an important consideration in the conservation evaluation process and the therefore should always be stated in conjunction with the IUCN classification. These species classifications were essential for comparison of conservation value between sites. This comprehensive approach was indispensable given the global importance of the water beetle fauna on Table Mountain (Perkins and Balfour-Browne 1994).

Data was presence/absence transformed using PRIMER v.4 (Clarke & Warwick 1994) utilising the Bray–Curtis coefficient on the transformed data. Hierarchical agglomerative clustering (HACA), using group-average linking, was executed on the data matrix. Ordination of samples by MDS was carried out, and stress values used to assess the reliability of the MDS ordination. Distinguishing taxa responsible for the similarity within group of sites and the dissimilarity between groups of sites were established using SIMPER (PRIMER 1994, Version 4).

Results

A simple assessment of the sampling techniques employed (Tables 2 & 3) resulted in marked differences in the species data. The number of species encountered would have been a mean of 44% lower had standard netting and elementary SASS been employed in the absence of supplementary specialist techniques. The mean incidence of endemic species and those of high conservation importance would have been respectively 53% and 50% lower. Across the data set this difference is further characterised by the rather serious, in terms of site comparability, differences between otherwise fauna rich sites and those with a poorer fauna, as exemplified by ihh and bhh (Table 3) where additional sampling would have had a minimal effect with only a 33% loss in endemic species encountered.

Given the preliminary nature of this study, the naturally small size of the upland streams and the resultant limited quantitative data, diversity indices have not been applied but rather the absolute frequencies considered adequate to indicate trends. In the upland sites the control (vrg) site and the Disa River above the reservoir (ahh) displayed the highest frequency of species, 9 and 11 respectively of which a remarkable proportion were endemic, 78% (vrg) and 82% (ahh), and of high conservation status 66% (vrg) and 73% (ahh). The endemic and endangered species were dominated by the hydraenids including the Table Mountain endemic *Prostethops nitens* (at ahh) and the Table Mountain proximal endemic *Prostethops megacephalus* (at vrg) and the new to science *Mesoceration* sp. cf. *abstrictum* (at ahh). The next most dominant taxa were the Elmidae with three species, of regionally endemic genera, found at both sites *Ctenelmis incerta*, *Elpidelmis capensis* and *Leielmis* sp. cf. *georyssoides*, the latter new to science. The Dryopidae were represented at both sites by further members of regional endemic genera *Rapnus raffrayi* and *Strina aurichalcea*. The primary diversity at these sites was held within these taxa, however frequency dominance was displayed by the Dytiscidae, primarily *Canthyporus hottentottus* and *Hydropeplus trimaculatus* (see below).

Table 2 Species composition, endemism and provisional conservation status across sites

Taxa	IUCN status	Regional endemism	vrg	ahh	ihh	bhh	dis	ori
Aulonogyrus (Afrogyrus) abdominalis (Aubé 1838)	LR(lc)							√
Aulonogyrus (Afrogyrus) marginatus (Aubé 1838)	LR(lc)							√
Aulonogyrus (Afrogyrus) sp. nov.	DD	CE					√	√
Canthyporus hottentotus (Gemminiger & Harcourt 1868)	LR(lc)	CE	√	√	√	√		√
Canthyporus exilis (Boheman 1848)	VU	CE		√				
Herophydrus obscurus Sharp 1882	LR(lc)	CE		√	√	√		
Hydropeplus trimaculatus (Laporte 1835)	VU	CE	√	√	√	√		
Mesoceration sp. nov. cf. *abstrictum* Perkins & Balfour-Browne 1994	CR	CE		√				
Prostethops nitens (Péringuey 1892)	CR	CE		√				
Prostethops megacephalus (Boheman 1851)	CR	CE	√					
Paracymus sp. nov.	DD	CE						√
Paracymus sp. indet. ♀			√					
Strina aurichalcea Redtenbacher 1867	VU	CE	√	√				
Rapnus raffrayi Grouvelle 1899	VU	CE	√	√				
Ctenelmis incerta (Grouvelle 1890)	LR(lc)		√	√				
Leielmis sp. nov. cf. georyssoides (Grouvelle 1890)	CR	CE	√	√				
Elpidelmis capensis (Grouvelle 1890)	VU	CE	√	√				√

Key: ahh = above, Hely-Hutchinson ihh = in Hely–Hutchinson, bhh = below Woodhead Reservoir, vrg = Valley of the Red Gods, dis = Disa River, ori = Original Disa Stream, CE = Cape endemic, LR(lc) = IUCN low risk (locally common), CR = IUCN Critically endangered, VU = IUCN Vulnerable, DD = IUCN Data Deficient

Table 3 Frequency and conservation status of species by site

Site	Full sample			Net samples only			% reduction in net only samples			General invertebrate abundance
	Freq. species	Freq. endemics	Freq. VU+	Freq. species	Freq. endemics	Freq. VU+	% Freq. species	% Freq. endemics	% Freq. VU+	
vrg	9	7	6	4	4	3	–55%	–42%	–50%	extremely abundant
ahh	11	9	8	6	6	4	–45%	–33%	–50%	extremely abundant
ihh	3	3	1	3	2	1	0%	–33%	0%	very low frequency
bhh	3	3	1	3	2	1	0%	–33%	0%	very low frequency
dis	1	1	1	0	0	0	–100%	–100%	–100%	low frequency
ori	6	4	3	2	1	0	–66%	–75%	–100%	low frequency
Mean % reduction by sampling with net only							–44%	–53%	–50%	

Key: ahh = above, Hely–Hutchinson, ihh = in Hely–Hutchinson, bhh = below Woodhead Reservoir, vrg = Valley of the Red Gods, dis = Disa River, ori = Original Disa Stream

The HACA analysis (Fig. 1) showed that the two sites above the reservoir (vrg, ahh) had a remarkable 76% similarity, and as illustrated in the MDS analysis (Fig. 2) form a distinct subgroup away from the within (ihh) and below reservoir (bhh) samples. The above and below reservoir groups were only 45% similar with SIMPER revealing the species in common as the Dytiscids *C. hottentottus*, *H. trimaculatus* and *Herophydrus obscurus*. The species absent from the reservoir and immediately below it can all be attributed to the dominant endemic and threatened taxa of the Hydraenidae, Elmidae and Dryopidae as discussed.

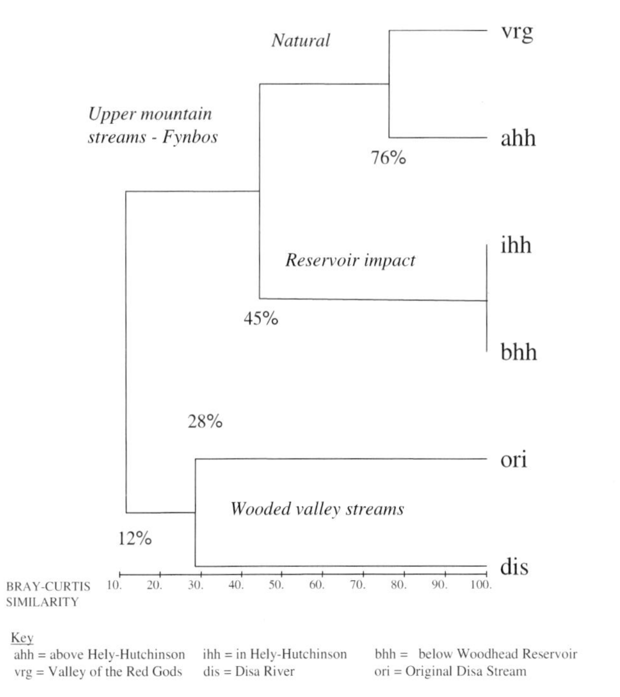

Fig. 1 HACA analysis of full samples between sites. *Key*: ahh = above, Hely–Hutchinson, ihh = in Hely–Hutchinson, bhh = below Woodhead Reservoir, vrg = Valley of the Red Gods, dis = Disa River, ori = Original Disa Stream

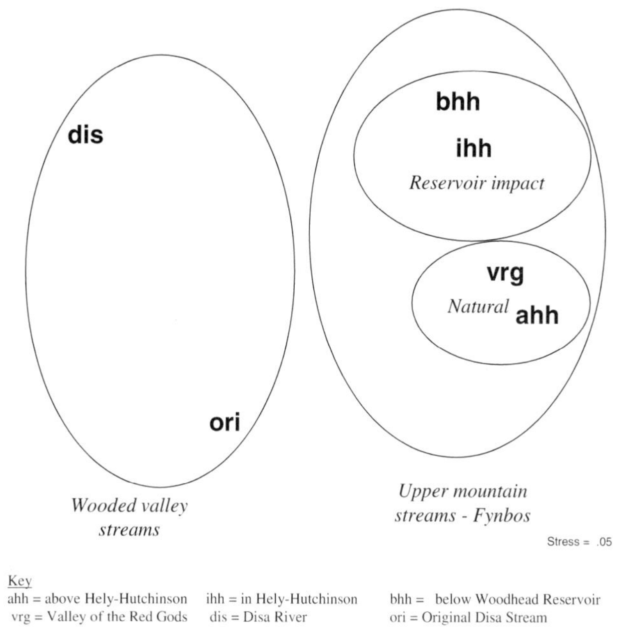

Fig. 2 MDS ordination analysis of full samples between sites. *Key*: ahh = above, Hely–Hutchinson, ihh = in Hely–Hutchinson, bhh = below Woodhead Reservoir, vrg = Valley of the Red Gods, dis = Disa River, ori = Original Disa Stream

In the reservoir the number of species encountered fell dramatically to 3, all Dytiscid species often found in the slow flowing riverine pools in the area. *C. hottentottus* has been captured in a wide variety of habitats from permanent and seasonal flowing waters to hypersaline coastal supralittoral pools (Turner 2005, 2006 pers. obs.) and they exceeded 200 in number in standard net samples above the reservoir (ahh) and in the control (vrg) whilst only present in very small numbers in the reservoir (ihh). The more specialist species of mountainous Cape habitats (Omer-Cooper 1965; pers. obs.) *H. trimaculatus* was also observed in large numbers (>100) at both above reservoir sites (vrg and ahh) whilst it was only represented by singletons in the reservoir itself. Both *C. hottentottus* and *H. trimaculatus* should be considered present in the reservoir because of dry season induced population pressure and subsequent inevitable dispersal from the natural riverine habitat above the reservoir. Had the reservoir (ihh) produced more than a few specimens of each this hypothesis would have been considered less viable. The presence of *H. obscurus* is perhaps less defined but it is known as a widespread species and may be encountered in a variety of habitats in low numbers (Biström and Nilsson 2002; pers. obs.). It therefore can be perceived as a generalist species endemic to the Western Cape. All three of the reservoir species were endemic producing an unusual 100% endemism for the site, of which only one was of high conservation status, *H. trimaculatus*. Below the reservoir (bhh) the species encountered were the same as in the reservoir with again 100% endemism, 33% (or 1 species) of conservation interest and only three species.

The lowland Disa River (dis) possessed the lowest diversity of all the sites sampled with only one species captured. Albeit this species is considered new to science and was also captured in the adjacent Original Disa Stream (ori) where considerably more species were encountered. Of the 6 species captured 4 (66%) were endemic and three (50%) of conservation interest (Table 2). The differences between the upland and lowland habitats were evident in the HACA and MDS analyses (Figs. 1 and 2). HACA analysis revealed that the upland (vrg, ahh, ihh, bhh) and lowland habitats (ori, dis) were only 12% similar with SIMPER determining that the species composition of the lowland rivers was very different to the unforested Fynbos surrounded upland waters with four species occurring only at the lower altitude, *Aulonogyrus abdominalis*, *Aulonogyrus marginatus*, *Afrogyrus* sp., *Paracymus* sp. and 11 only present at the higher altitudes: *Canthyporus exilis*, *H. obscurus*, *H. trimaculatus*, *Mesoceration sp. cf. abstrictum*, *Prostethops nitens*, *Prostethops megacephalus*, *Paracymus* sp. indet. *female*, *S. aurichalcea*, *R. raffrayi*, *Ctenelmis incerta*, *Leielmis sp. cf. georyssoides* as exemplified in Table 2. This dissimilarity created two clear MDS groupings clearly separating the lowland habitat altitude c. 150 m from that of the upland altitudes c. 720–780 m. Within the lowland grouping the two sites were only 28% similar (Fig. 1).

Casual observations of the other aquatic invertebrates in the samples appeared to indicate similar trends to those displayed by the water beetles between sites. This was particularly true for the Crustacea where they occurred in profusion above the reservoir (ahh) and in the Valley of the Red Gods (vrg) in contrast to the comparatively very small number of individuals below (bhh) and in the reservoir (ihh).

Discussion

This survey examined the distribution of seventeen species between study sites on Table Mountain. Four of these species were considered new to science and will be the subject of further investigations, whilst only one species was indeterminate. A remarkable 77% of the species encountered were Cape endemics and 65% were of provisionally IUCN internationally vulnerable or higher; this included the data deficient species considered as new. This study has provided important records of some Table Mountain endemic and proximal endemic species as well as providing a brief insight into the complex, but nonetheless clear, impacts of reservoir construction and altitudinal habitat differentiation.

Given that this research was intended as a preliminary study and relatively low sampling effort expended the results were remarkably definitive. The impact of the reservoirs was devastating to all the members of the taxa displaying high levels of endemism and conservation status, namely the Hydraenidae, Elmidae and Dryopidae. A 73% loss in species diversity was experienced between the Disa River above the Hely–Hutchinson reservoir when compared to, in, or below the reservoir. The loss comprised of 7 (64%) species with IUCN provisionally internationally vulnerable status, including data deficient species and 8 (73%) Cape endemic species. Casual observation suggested that this loss was also reflected by other aquatic invertebrate taxa. This loss was not considered an artifact of the relatively small change in altitude and is supported by preliminary results from an adjacent site, at a lower altitude of c. 680 m, Kasteelspoort, where species diversity is higher than at either of the control at the Valley of the Red Gods or the Disa River. Dam construction was included by Foster (pers. comm.) as a serious threat to water beetle species and reiterated for the South African fauna by Turner (in prep. a, b). The potential impact of reservoirs, as inferred by work on other taxa (i.e. Dadzie 1980; Beadle 1981; O'Keeffe 1985; Jackson et al. 1988; Davies and Day 1998; Gourène et al. 1999; Da Costa et al. 2000; Skelton 2001; Gratwicke 2003) has been confirmed here for the water beetle fauna. The mechanism for this dramatic change in species composition may partially be attributed to the change from a riverine to lacustrine habitat given current knowledge of the ecological niche occupied by the endemic Hydraenidae, Elmidae and Dryopidae species (Perkins and Balfour-Browne 1994;

pers. obs.). However, this only explains the absence of the species in the Hely–Hutchinson Reservoir not the continued absence of the species below the reservoir that may be attributable to more subtle changes in water chemistry, temperature or hydrological cycle. The Disa River feeding the Hely–Hutchinson Reservoir was found to contain a fauna more diverse than the control site and these two sites were distinct from the other study localities and classified as natural upland habitat for Table Mountain. The reservoir and its outlet displayed the species composition characteristic of natural upland habitat but only with respect to the sole occupants—the dytiscid species. These natural and reservoir impacted upland aggregations were found to be 45% similar, whilst the within group similarities were 100% for reservoir impacted and 76% for the more diverse natural grouping. There was a distinct 88% difference in species composition between the upland (720–780 m) and lowland (150 m) habitats on Table Mountain, and unlike the upland habitats the similarity between lowland site species composition was low at only 28%. The absence of additional lowland sample locations could have been a contributory factor in the low resolution of this aggregation.

The issue of diverse sampling techniques for bioinventory had been raised by Bratton (2001), Foster (pers. comm.), and in regard to the South African fauna by Perkins (2004 a, b; 2005 a, b) and Turner (2004, 2005, 2006). The simplistic assessment of sampling techniques (Table 3) illustrated the importance of a Balanced Scorecard approach to comparative and bioinventory studies where a combination of established quantitative methodology and specialist techniques must be employed. Sole employment of standard netting would have resulted in remarkably less endemic (53%) and conservation important (50%) species included in the study and inferred that the comparative results would have been severely compromised between sites. This use of standard netting positively selected for free swimming Dytiscidae, in this instance characteristic of less specialist environments, (i.e. *Canthyporus* Zimmermann 1919, *Herophydrus* Sharp 1882 and *Hydropeplus* Sharp 1882) and the generalist substrate occupying species belonging to genera relatively easily disturbed from their refugia (i.e. *Rapnus* Grouvelle 1899, *Strina* Redtenbacher 1867 and *Ctenelmis* Delève 1964). In particular the lack of endemic specialist genera in the standard samples (i.e. *Mesoceration* Janssens 1967, *Prostethops* Waterhouse 1879 and *Leielmis* Delève 1964) would have severely impaired the results although initial indications appear to show that comparative differences between sites would have been present but not as markedly distinct. Interestingly the recent expansion of the SSAS methodology (Dickens and Graham 2002) reflected recognition of this shortfall in resolution, but full expansion of the methodology was prevented by the requirement for national standardization and utilization across a broad skill base and the consideration for encompassing a broad spectrum of taxa.

Care must be taken when considering the emphasis of conservation efforts in the light of survey results and extractable statistics. This was exemplified by the presence of 100% endemism in the species found immediately below and in the Hely-Hutchinson Reservoir, the highest levels recorded in this study. This superficially suggested that reservoirs were beneficial. In isolation this statistic is impressive, but in reality the samples comprised of only three species, all of which occur at a minimum of three sites in the survey and two species of which are widespread in the region. The low community diversity and low conservation status frequencies compared to the other pristine sites (ahh, vrg), shown in Table 2, place this isolated statistic in perspective as highly subjective and essentially irrelevant to conservation decisions other than to confirm the negative effect of reservoir construction. This scenario stresses the importance of the requirement for specialist examination of base line data for conservation and site evaluation during the consultation process and prior to the approval of development plans, site management or conservation strategy implementation.

This survey conclusively presented the extremely destructive impact of artificial inundation to the conservation priority water beetle species of Table Mountain. This negative impact is likely to be applicable to the remainder of the region and further afield. Conservation assessment of survey localities must be shown to have utilised full bioinventory methodologies (Turner in prep b) and not exclusively employ standard netting techniques (i.e. SASS). The evaluation of species data for conservation, environmental impact assessment (EIA) or habitat management purposes must include the balanced consideration of at least recent survey data, the provisional IUCN conservation status and level of endemism. Environmental data must be reviewed by a specialist to facilitate the balanced and appropriate implementation of decisions. This approach can be referred to as the Balanced Scorecard Conservation Assessment. It is recommended that any EIA should include water beetle community evaluation before any dam construction, hydrological alteration or management plan implementation with a potential influence on natural water flow. A comprehensive survey of the water beetles of Table Mountain and the Cape Peninsula, throughout the year or at least during the dry and the wet seasons, is recommended with a view to attaining at least a minimal database on the species in the area, their distributions, ecological requirements and conservation statuses. Table Mountain and the Cape Peninsula remain globally unparalleled in their biodiversity and species conservation importance despite the continued negative influence of the growing conurbation of Cape Town. It is imperative that the utmost is done to continue to conserve this unique example of biodiversity under siege from a demanding conurbation.

Acknowledgements I would like to thank Mr. Hennie Homan (Orangekloof) and Mr. Andre du Plessis (Newlands) of Table Mountain Reserve, Cape Town for their invaluable assistance and enthusiasm in the field and approval for study in their management areas. Ms. Julia Wood of the City Engineer's Department, City of Capetown for her and the department's invaluable assistance in obtaining the relevant study permits. Mrs. Liebenberg of Cape Nature Conservation for her and the organisation's assistance in providing the relevant study and export permits. Roy Ernstein of the Cape of Good Hope Nature Reserve for his assistance and advice on permit applications. In the UK, Dr. David Bilton of Plymouth University, Darren J. Mann of the Hope Entomological Collections, Oxford Museum of Natural History and Maxwell V. L. Barclay, Entomology Department, Natural History Museum, London for their useful discussion and comment on material, Prof. Garth Foster, Aberdeen University, UK, Prof. Michael Samways of Stellenbosch University, South Africa and Prof. Tim New of La Trobe University, Australia, for their enthusiastic approach to this paper and encouragement to publish.

References

Bassen MS, Van Niekerk PH, Van Rooyen JA (1997) Overview of water resources availability and utilization in South Africa. Department of Water Affairs and Forestry Pretoria, South Africa 72 pp

Beadle LC (1981) The inland waters of tropical Africa. Longman Group Ltd, USA ISBN0582463416, pp 475

Biström O, Nilsson AN (2002) Herophydrus Sharp: Cladistic analysis, taxonomic revision of the African species and world checklist. Koleopterologische Rundschau 72:15–111

Branch WR (1989) Endangered reptiles and amphibians in South Africa. Afr Wildlife 43:195–201

Bratton JH (2001) A comparison of the efficiency of pond net and sieve for catching aquatic beetles and aquatic bugs over the course of a year in a grazing levels ditch in Malltraeth Marsh, Anglesey, Wales. Countryside Council for Wales, Natural Science Report No. 01/5/1

City of Cape Town (CCT) (2001) Bulk water supply infrastructure: Background information for WSDP. www.capetown.gov.za/water/wsdp/documents/wsdp-report-pdf.pdf

Chutter FM (1994) The rapid biological assessment of streams and river water quality by means of macroinvertebrate communities in South Africa. In: Uys MC (ed) Classification of rivers and environmental health indicators. Water Research Commission Report No TT 63/94, South Africa, pp 217–234

Clarke KR, Warwick RM (1994) Change in marine communities, Primer version 4 instruction manual. Plymouth University, UK

Da Costa KS, Gourène G, Tito De Morais L, Thys Van Den Audenaerde DFE (2000) Caractérisation des peuplements ichtyologiques de deux fleuves côtiers ouest-africains soumis à des aménagements hydroagricoles et hydroélectriques. Vie et Milieu 50:65–77

Dadzie S (1980) Recent changes in the fishery of a new tropical man-made lake, Lake Kamburu (Kenya). J Fish Biol 16:361–367

Dallas H (2000) Ecological reference conditions for riverine macroinvertebrates and the River Health Programme, South Africa. 1st WARFSA/WaterNet Symposium: Sustainable Use of Water Resources; Maputo, 1–2 November 2000

Davies BR, Day J (1998) Vanishing Waters. University of Cape Town Press, Cape Town

Day J (2000). Biomonitoring: appropriate technology for the 21st century. 1st WARFSA/WaterNet Symposium: Sustainable Use of Water Resources, Maputo, 1–2 November 2000

Dickens CWS, Graham PM (2002) The South African Scoring System (SASS) version 5 rapid bioassessment method for rivers. Afr J Aquat Sci 27:1–10

Gourène G, Teugels GG, Hugueny B, Thys Van Den Audenaerde DFE (1999) Evaluation de la diversité ichtyologique d'un basin ouest Africain après la construction d'un barrage. Cybium 23:147–160

Gratwicke B, Marshall BE, Nhiwatiwa T (2003) The distribution and relative abundance of stream fishes in the upper Manyame River, Zimbabwe, in relation to land use, pollution and exotic predators. Afr J Aquat Sci 28(1):25–34

Gueorguiev VB (1967) Coleoptera: Haliplidae. In Hanstrom B, Brinck P, Rudebeck G (eds) South African Animal Life: results of the Lund Univ. Expedition in 1950–1951, vol. 13. Swedish Nat. Sci. Res. Council Stockholm, pp 9–14

Guignot F (1959) Revision des Hydrocanthares d'Afrique (Coleoptera Dytiscoidea), Part 1. Ann Mus Roy Congo Belgique (ser 8) Sci Zool 70:1–316

Jackson PBN, Marshall BE, Paugy D (1988) Fish communities in man-made lakes. In Lévêque C, Bruton MN, Ssentongo GW (eds) Biology and ecology of African freshwater fishes. ORSTOM, Paris, pp 325–350

Komarek A (2004) Taxonomic revision of *Anacaena* Thomson 1859 I. Afrotropical species (Coleoptera: Hydrophilidae). Koleopterologische Rundschau 74:303–349

Mace GM, Stuart SN (1994) Draft IUCN Red List Categories, Version 2.2. Species 21–22:13–24

Madikizela BR, Dye AH, O'Keeffe JH (2001) Water quality and faunal studies in the Umzimvubu catchment, Eastern Cape, with particular emphasis on species as indicators of environmental change. WRC Report No. 716/1/01, Water Research Commission, Pretoria

Madikizela BR, Dye AH (2003) Community composition and distribution of macroinvertebrates in the Umzimvubu River, South Africa: a pre-impoundment study. Afr J Aquat Sci 28(2):137–149

Mountain Club of South Africa (MCSA) (1993) Approved paths on Table Mountain. Cape Town Section, Mountain Club of South Africa, South Africa, Scale 1:12500

O'Keeffe JH (1985) The blackfly problem in the Great Fish River.The Naturalist 29(3):3–7

Omer-Cooper J (1962) The distribution of the Cape Province Dytiscidae and speculations as to their possible origins. Ann Cape Province Museum II:292–296

Omer-Cooper J (1965) Coleoptera Dytiscidae. A review of the Dytiscidae of Southern Africa. South Afr Life (Results of the Lund University Expedition in 1950–51) 11:59–214

Pauw A, Johnson S (2002) Table Mountain: a natural history. Fernwood Press, South Africa, ISBN1874950431, pp 160

Perkins PD, Balfour-Browne J (1994) A contribution to the taxonomy of aquatic and humicolous beetles of the family Hydraenidae in southern Africa. Fieldiana Zoology (New Series). Field Museum of Natural History, ISSN00150754, 77:1–159

Perkins PD (2004a) A revision of the South African endemic water beetle genus *Pneuminion* Perkins (Coleoptera: Hydraenidae). Zootaxa 538:1–38 ISSN11755334

Perkins PD (2004b) A revision of the South African endemic humicolous beetle genus *Nucleotops* Perkins & Balfour-Browne. Zootaxa 559:1–12

Perkins PD (2005a) A revision of the South African endemic humicolous beetle genus *Discozantaena* Perkins & Balfour-Browne. Zootaxa 915:1–48

Perkins PD (2005b) A revision of the African hygropetric genus *Coelometopon* and description of *Oomtelecopon* new genus (Coleoptera; Hydraenidae). Zootaxa 949 103pp ISBN-187735497X

Ribera I, Beutel RG, Balke M, Vogler AP (2002) Discovery of Aspidytidae, a new family of aquatic Coleoptera. Proc R Soc Lond B 269:2351–2356

Skelton P (2001) A complete guide to the freshwater fishes of Southern Africa. Struik Publishers, South Africa, ISBN1868726436, pp 395

Slingsby P (2004) Table Mountain—the map. Baardskeerder, South Africa, ISBN191990073X, Scale 1:20,000

Turner CR (2004) Towards conserving the aquatic Coleoptera of southern Africa. IUCN SAISG, Colophon 4:11–16

Turner CR (2005) Ochthebius (s.str.) capicola (Péringuey, 1892) (Coleoptera) rediscovered in the Cape Peninsula, South Africa. Afr J Aquat Sci 29(2):271–273

Turner CR (2006) Recent advances in the study of the aquatic Coleoptera of Southern Africa, Colophon 6: in prep

Turner CR (in prep. a) The water beetles (Coleoptera) of South Africa, Lesotho and Swaziland—a provisional international conservation status review

Turner CR (in prep. b) Chapter 2: Studying the aquatic Coleoptera—a practical guide to bioinventory with special reference to southern Africa. IN: Guides to the Freshwater Invertebrates of Southern Africa: Volume 10 Coleoptera, Water Research Commission of South Africa

Uys MU, Goetsch MA, O'Keeffe JH (1996) Ecological indicators, a review and recommendations. National Biomonitoring Programme for Riverine Ecosystems Report Series No 4. Institute for Water Quality Studies, Department of Water Affairs and Forestry, Pretoria, South Africa, 93 pp

J Insect Conserv (2007) 11:85–94
DOI 10.1007/s10841-006-9021-1

BEETLE CONSERVATION

The decline of native coccinellids (Coleoptera: Coccinellidae) in the United States and Canada

Jason P. Harmon · Erin Stephens · John Losey

Received: 9 February 2006 / Accepted: 7 April 2006 / Published online: 2 November 2006

Abstract Reviewing published coccinellid surveys we found that the number of adventive species has increased steadily over the last century while the average proportion of native individuals has remained fairly constant until 1987 followed by a rapid decrease between 1987 and 2006. Seven long-term studies indicated that the total density of coccinellids increased by an average of 14% following establishment of adventive species, but this increase was not significant and in 4 of 7 cases the total density of coccinellids actually decreased following establishment. Similarly, no significant difference was found in comparisons of diversity across all studies. These results illustrate that even with multiple long-term data sets it is currently difficult to make any general conclusions regarding the impact adventive coccinellids have had on native coccinellid assemblages. However, it is clear that specific systems and species have seen major shifts in recent years. For example, adventives have become the dominant species in a third of the assemblages where they are found. Focusing on two formerly common native species, *Adalia bipunctata* and *Coccinella novemnotata*, we show they have become rare in their former ranges and discuss potential explanations for this phenomenon.

Keywords *Adalia* · *Coccinella* · Adventive species · Ladybirds · Aphids

J. P. Harmon (✉)
Department of Zoology, University of Wisconsin, 430 Lincoln Dr., Madison, WI 53706, USA
e-mail: jharmon@wisc.edu

E. Stephens · J. Losey
Department of Entomology, Cornell University, Ithaca, NY 14853, USA

The importance of coccinellid diversity and abundance

Coccinellids, known as ladybugs, ladybeetles, or ladybird beetles are one of the most common and easily recognizable invertebrate components of almost every terrestrial ecosystem in the US and Canada (Gordon 1985). Species in this family are so ubiquitous and yet so sensitive to environmental conditions that they have been proposed as indicator species (Iperti 1999).

This species complex also contributes to the regulation of many soft-bodied insects, especially those in the order Homoptera, and is valuable in controlling the egg and larval stages of other insects (reviewed in Hodek and Honěk 1996). The services that these predators provide are so well recognized and respected that pest management practices are often designed to take advantage of their pest suppression potential. For example, in alfalfa the decision to take action to suppress aphid populations is based on both aphid and coccinellid densities (e.g., Knowles 2006). Pesticides are not applied even if aphid densities exceed what would be economically damaging levels if the density of coccinellids is high enough to suppress them.

Given their potential to control pest species, many programs have tried to supplement extant populations or introduce new species. Purchasing and releasing natural enemies to augment biological control is a potentially valuable and continually growing practice with coccinellids being one of the most important groups used (Cranshaw et al. 1996). Vast resources have also been expended to introduce and permanently establish coccinellid species that are not native to the Nearctic region. One of the first successes with classical biological in the US involved importing the vedalia beetle, *Rodalia cardinalis*, from Australia to suppress the cottony cushion scale, *Icerya purchasi*, in California. The entire cost of this project is estimated at $2,000 (Pedigo and Rice 2006), approximately $41,000 in 2005 adjusting for inflation. This figure is probably below the average cost of subsequent introductions given the increased level of pre-release research that is usually done. Taking the adjusted figure as an average cost for the 179 coccinellid introductions into North America (Gordon 1985) the total cost of these efforts exceeds 7 million dollars.

Given their charisma and acknowledged importance, it is not surprising that numerous studies have examined the composition of this complex and a smaller but no less important body of literature has related their density and diversity to ecological function. Two clear emergent properties are that coccinellid species vary widely in the level of suppression they exert on various prey species (reviewed in Hodek and Honěk 1996) and that species vary widely in their response to environmental changes (Iperti 1999; Bazzocchi et al. 2004). Thus, long-term regional shifts in species composition may have important implications for the functioning of this complex and its response to environmental changes. In this paper we draw on published surveys to examine the density and composition of aphidophagous coccinellids in the US and Canada over the past hundred years.

Materials & methods

We reviewed the published literature for surveys of adult aphidophagous coccinellids in natural and managed ecosystems throughout the United States and Canada. Despite the importance of a number of coccidophagous coccinellids for biological control in North America, e.g., *Rodolia cardinalis* (DeBach 1964), we focus here on aphidophagous species due to their ubiquitous nature, their dominance in the literature, and their importance in recent purposeful and accidental invasions. Studies were included if there were at least 20 naturally occurring individuals collected, the proportion of native and adventive species was easily determined from the paper, and >95% of individuals were identified to species. Only a very small number of studies included larval information, so for consistency we only used information on adult coccinellids. We extracted multiple data sets if collections were made in distinct large-scale habitats (e.g., two different crops) or if multi-year studies had distinct sampling periods with considerable differences (e.g., before and after an invasion). We determined the average coccinellid assemblage for all other multiyear surveys by calculating the proportion of each species in the assemblage for each year and then averaging across years. For analyses over time we used the midpoint date (rounded up) for these multiyear surveys. We also determined average assemblages in the same manner when surveys took place across multiple but similar habitats (e.g., multiple fields of the same crop). Note that in a few instances (e.g., Putnam 1964) information was not available to calculate the average assemblage in this way and we were forced to use an average assemblage as determined by the author. These criteria resulted in information on 71 coccinellid assemblages from 36 references (Ewing 1914; Fluke 1925; Dobzhansky 1935; Fenton and Howell 1955; Godarzy and Davis 1956; Smith 1958; Putnam 1964; Day 1965; Gagne and Martin 1968; Smith 1971; Wheeler 1971; Watve and Clower 1976; Angalet et al. 1979; Turnock and Turnock 1979; Lee 1980; Dowell and Cherry 1981; Mareida et al. 1992; Elliott et al. 1996; LaMana and Miller 1996; Colunga-Garcia et al. 1997; Hoffmann et al. 1997; Brown and Miller 1998; Colunga-Garcia and Gage 1998; Boiteau et al. 1999; Cormier et al. 2000; Hesler et al. 2000; Wright and DeVries 2000; Wold et al. 2001; Bosque-Perez et al. 2002; Stephens 2002; Brown 2003; Turnock et al. 2003; Alyokhin and Sewell 2004; Evans 2004; Hesler et al. 2004; Musser et al. 2004). We further calculated measures of species richness, the Berger–Parker Dominance Index, as well as Simpson's D and Shannon's H diversity indices for the 62 data sets in which all collected individuals were identified to species (Magurran 1988). Since both diversity indices gave very similar results, we only report Simpson's D here.

Results and discussion

Establishment and increase of adventive species

It is difficult to discuss the status of native species without also considering the data on adventive species. The evidence for a causal relationship between the establishment of adventive species and the decline of native species is by no means conclusive (as we discuss below). However, they have certainly changed the composition of this complex by their presence alone.

At least 179 coccinellid species have been introduced deliberately or inadvertently and 27 have become established in the US and Canada (Fig. 1) (Gordon 1985; Gordon and Vandenberg 1991). It appears that although the establishment of many species has been confirmed, very few have grown to levels where they are commonly found in published surveys.

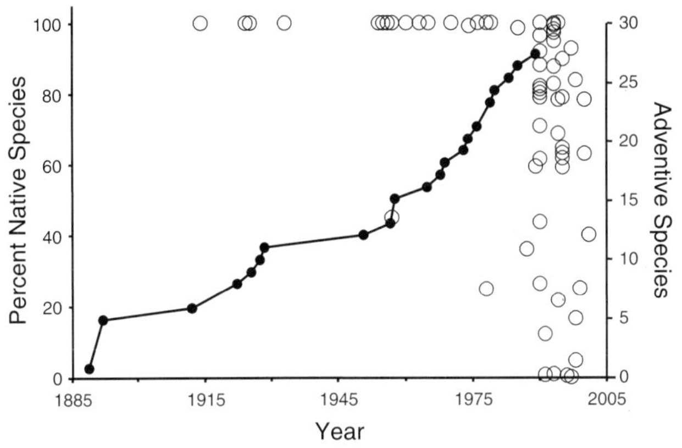

Fig. 1 The cumulative number of adventive species established in the United States and Canada (line) and the percent native species found in surveys of coccinellid assemblages (circles) over time

Only six adventive species were reported in studies of full coccinellid assemblages. Of these, *Coccinella septempunctata* was reported the most often, followed by *Harmonia axyridis*, *Propylea quattuordecimpunctata*, *Coccinella undecimpunctata*, and single occurrences of *Exochomus quadripustulatus* and *Hippodamia variegata*. This is almost certainly due, at least in part, to a sampling bias since most of our data come from surveys in cultivated herbaceous plants (e.g., alfalfa, corn).

Overall, the number of adventive species found within a coccinellid assemblage varied from zero to three with more adventive species being found in more recent studies ($F_{1,61}$=31.5, P<0.001), especially those after 1985. At least two of the most common adventive species, *C. undecimpunctata* and *C. septempunctata*, show a pattern of slow increase after establishment followed by a period of rapid growth. More surprising is that for both species this period of rapid growth and rise to prominence was relatively short-lived. *Coccinella undecimpunctata* was the most common species found in potatoes in New York in 1957 (Day 1965) and natural areas in New Jersey in 1978 (Angalet et al. 1979), but was not reported in earlier or later surveys. *Coccinella septempunctata* was the most common species in many surveys in the 1980s but was frequently superseded by another adventive species, *H. axyridis*, in the 1990s. It seems plausible that *H. axyridis*, currently the most common adventive coccinellid in the US and Canada, could undergo a similar decline; however it is unclear which species would replace it since no new species have established in almost 20 years and several key natives have declined to very low densities.

Proportional change in the coccinellid complex

While the number of adventive species has increased over the past 100 years the proportion of native individuals has decreased markedly (Fig. 1). The increase in the number of adventive species has been steady, nearly linear, yet native individuals stayed very close to 100% of the assemblage (with two notable exceptions) until the mid 1980s when assemblages began to vary tremendously in the reported proportion of natives. From 1914 to 1985 native individuals averaged 95.0%±4.0 (1 S.E.) while from 1987 to 2001 natives were significantly lower at 67.5%±4.0 (t_{69}=4.48, P < 0.001). This pattern suggests that if adventive species are the cause of the decline of native species then either the total impact of all these adventive species reached some critical level, or that a subset of the more recently introduced species are causing the bulk of the impact. Based on their extremely high densities and relatively recent introductions, *C. septempunctata* and *H. axyridis* may be responsible for much of the impact to these surveyed native assemblages.

While this proportional decline is striking, care must be taken to not over-interpret these results. A decline in the proportion of natives can result from a decrease in native density or by simply increasing total coccinellid density (natives + adventives) without increasing the density of natives (discussed below). Thus, a complete assessment requires interpreting data on both proportions of native species and their densities.

Numerical change in the coccinellid complex

Most of our collected data sets cannot be used to investigate changes in coccinellid densities. The wide variation in densities across time, space, habitat, and collecting method made it impossible to use most of our data to look for causality by making meaningful comparisons of density across specified time periods (e.g., before and after establishment of an adventive species). Fortunately, we found seven data sets from five studies that measured coccinellid abundances in a single system and with the same method over multiple years that encompassed the establishment of an adventive species (see Table 1). These examples exhibit mixed results in terms of the effect of adventive species on coccinellid densities. Overall the total density of coccinellids increased by an average of 14% (14±21) following establishment. However, the tremendous variation in this study overwhelms any attempt at a single average; in 4 of 7 cases the total density of coccinellids decreased following establishment. This high variability led to no overall difference in total coccinellid density before compared to after invasion (t_5=0.66, P=0.53).

Table 1 Multi-year surveys of coccinellid populations

	Habitat	Units	Total before introduction	Total after introduction	Total after introduction minus focal invader[a]	Adventive Coccinellid
Elliott et al. (1996)	Alfalfa	Ave per 50 sweeps	2.30 1973–1985	3.31 1988–1992	2.69	*Coccinella septempunctata*
	Corn	Ave Per 15-min search	18.4 1973–1985	15.8 1988–1992	14.5	*Coccinella septempunctata*
	Grains	Ave per 50 sweeps	2.90 1973–1985	3.84 1988–1992	3.08	*Coccinella septempunctata*
Colunga-Garcia et al. (1998)	Varied	Ave/210 traps/season	3082.3 1989–1993	2497.5 1994–1998	2112.5	*Harmonia axyridis*
Turnock et al. (2003)	Alfalfa	# Collected (Averaged Per Year)	209.4 1983–1987	459 1988–2001	367.2	*Coccinella septempunctata*
Alyokhin and Sewell (2004)	Potato	Ave per sampled plant	0.0264 1971–1978	0.0212 1980–1995	0.0073	*Coccinella septempunctata*
Evans (2004)	Alfalfa	Ave # per 15 sweeps	4.23 1992–1994	2.34 1998–2001	0.61	*Coccinella septempunctata*

Columns contrast: (1) the total coccinellid populations before and after the arrival and/or establishment of the focal adventive species, and (2) the density of species found before invasion with the density of those same species after invasion

[a]Column is the density of the native coccinellid assemblage with the exception of Colunga-Garcia which includes *C. septempunctata* before and after the introduction of *H. axyridis*

We see a similar trend when looking at only the densities of native coccinellids. In the three studies where total coccinellid density increased, native density also increased following establishment of the adventive species and in the four studies where total density decreased native density decreased as well. Overall the density of native species declined by 16% (16±21) but again there was considerable variation and the difference between before and after establishment was not significant (t_5=0.77, P=0.47).

These studies illustrate that even with multiple long-term data sets it is difficult to make any general conclusions regarding the impact of adventive species establishment on the density of native species or the total density of the coccinellid complex. However, it is clear that in at least some specific cases both the density of native coccinellids and the size of the total coccinellid population was different before compared to after the arrival of an adventive coccinellid (Table 1). However, the direction, magnitude, and significance of such changes varied greatly across studies. Hopefully the discovery of more data will allow us to make firmer conclusions.

Richness, diversity and dominance of native species

In contrast to the low number of adventive species represented in our data set, we found 47 species of native coccinellids with the number of native species found in a single assemblage varying from 0 to 22. The number of native coccinellids found within a given assemblage did not seem to change over time ($F_{1,61}$=0.13, P=0.72), nor was there any relationship between the number of adventive species found in a assemblage and the number of native species ($F_{1,61}$=0.74, P=0.39). Taken together, these results suggest that there is no current evidence for an overall effect of adventive coccinellids on native species richness within a given assemblage.

The diversity of coccinellid assemblages with adventive species was only slightly higher than assemblages with all native species (Simpson's D: 2.67±0.16 at least one adventive species vs. 2.47±0.26 all natives, t_{61}=0.68, P=0.50). However, apart from any biological interactions, we would expect diversity to increase with the addition of adventive species. Therefore in addition to the total diversity of all coccinellids as reported above, we also calculated the diversity of only native coccinellids. In assemblages where adventives were found, the diversity of native coccinellids was of course lower than the total diversity (natives + adventives) of coccinellids in those systems (Simpson's D: 2.28±0.15), but they were still not significantly different from the diversity of assemblages with all native species (t_{60}=0.68, P=0.50). This again suggests that we have no evidence that the presence of adventive species has had a significant negative effect on the diversity of individual native coccinellid assemblages.

Adventive species could influence the composition of coccinellids without having a significant effect on a diversity index by replacing a dominant native species with a dominant adventive species. The coccinellid assemblages of our data set were often dominated by one or a small number of species. For example, on average the single most dominant species accounted for almost 60% of the entire assemblage (Average Berger–Parker's dominance index d=0.59). Within assemblages with adventive species, a native was still more likely to be the dominant species (30 vs. 14), but there was no difference in the magnitude of dominance between native dominated and adventive dominated systems (d=0.55±0.03 vs. 0.60±0.04, t_{42}=0.96, P=0.34). Even though native species are more likely to dominate, the shift from almost every assemblage being dominated by native species to almost a third being dominated by adventive species could be very important for the functioning and long-term stability of these assemblages. The dominance of adventive species may expand if adventive species such as *H. axyridis* continue to increase and key native species continue to decline.

A tale of two species: native species in rapid decline

While several authors have suggested that adventive coccinellid species are a major cause for a decline in specific native species (Wheeler and Hoebeke 1995; Ellis et al. 1999; Wise et al. 2001) we cannot support this as a general claim across all coccinellid assemblages. Similarly, Brown (2003) also did not find an impact of *H. axyridis* on the coccinellid complex when looking at original data and a review of studies in apples. Despite the lack of a general decline in natives, it is clear that several formerly common native coccinellid species are now quite rare. In this section we provide some detail on two species, *C. novemnotata* and *A. bipunctata*, that have experienced noted and rapid declines (Obrycki et al. 2000).

Adalia bipunctata

Evidence for decline

A number of coccinellid surveys in agricultural systems have indicated a decline in populations of *A. bipunctata* over the past 10–20 years (Table 2). Some of these most recent surveys indicate *A. bipunctata* is near or below the threshold of detection in places it had regularly inhabited. While these studies on the whole are limited spatially, temporally, and in the types of ecosystems surveyed, they present a consistent picture of decline. Most of these studies have discussed *A. bipunctata* decline in connection with the arrival of an adventive coccinellid. The first prominent report of a decline in *A. bipunctata* found it to be 20 times less abundant after the arrival of *C. septempunctata* (Elliott et al. 1996). Most studies since have shown a negative relationship between *A. bipunctata* populations and *H. axyridis*, but some of these studies have gone further to suggest that *A. bipunctata* abundance may had been declining even prior to *H. axyridis* arrival (Brown and Miller 1998; Colunga-Garcia and Gage 1998; Boiteau et al. 1999).

Potential reasons for decline

The strong temporal correlation between *A. bipunctata*'s decline and the emergence of *C. septempunctata* and *H. axyridis* is suggestive, yet sampling studies cannot demonstrate causality and we have little direct evidence of their interactions in North America. One method of evaluating potential causal factors is to assess their likelihood given studies of coccinellid ecology (e.g., Obrycki et al. 2000). Here, we briefly discuss some factors as they relate specifically to *A. bipunctata*.

The presence of *C. septempunctata* was the dominant explanatory factor for the observation of Elliott et al. (1996) and could explain the observed declines in *A. bipunctata* before *H. axyridis* was observed (Table 2). However, in a review of the ecology of *A. bipunctata* and *C. septempunctata*, Obrycki et al. (2000) concluded that the extent of their interactions should have been limited by spatial and temporal differences in habitat use. Given *A. bipunctata*'s seasonal changes in habitat use (reviewed in Hodek and Honěk 1996), one possible explanation for this apparent discrepancy is a strong source-sink dynamic across the landscape (e.g., Rosenheim 2001). Preferred, wooded habitats may act as source habitats for *A. bipunctata* while the sampled cultivated crops may act as sinks. If so, a habitat generalist such as *C. septempunctata* may generate greater negative effects in areas such as eastern South Dakota where wooded habitats are scarce (Obrycki et al. 2000) than in areas with more of *A. bipunctata*'s preferred habitat.

The ecology of *H. axyridis*, however, suggests that it could have a stronger effect on *A. bipunctata*. *Harmonia axyridis*, like *A. bipunctata* has a broad, polyphagous diet which includes many aphid species, and both use a broad range of habitats while preferring arboreal systems (Hodek and Honěk 1996). Egg and larval cannibalism appears to be important in *H. axyridis* population dynamics (Koch 2003), can be important in coccinellid interactions (Schellhorn 1998), and may be an important part of the interaction between *H. axyridis* and *A. bipunctata*. While *A. bipunctata* seems to have an advantage against *C. septempunctata*

Table 2 Surveys of *A. bipunctata* populations

	Habitat	Sample	Units	Before introduction	After introduction	Adventive Coccinellids
Elliott et al. (1996)	Corn—SD USA	Visual search	# / 15 min search	1973–1985 0.04±0.01	1988–1992 0.002±0.002	*Coccinella septempunctata* (later)
Colunga-Garcia and Gage (1998)	Agricultural landscape—MI USA	Sticky traps	Adults/210 traps/season	1989 115.3 1990 228.4 1991 36.5 1992 31.2 1993 16.6	1994 25.7 1995 10.2 1996 1.2 1997 0.2 1998 1.3	*Coccinella septempunctata, Harmonia axyridis* (later)
Brown and Miller (1998)	Apple trees—WV USA	Visual search	Percent of all ladybirds	1989–1990 9.1% 1991 2.3% 1993 1.7%	1994–1996 0%	*Coccinella septempunctata, Harmonia axyridis* (later)
Boiteau et al. (1999)	Agricultural landscape—NB Canada	Sticky traps (tower)	Total per summer	1992 362 1993 69 1994 51	1995 19	*Coccinella septempunctata, Harmonia axyridis* (later)
Cormier et al. (2000)	Herbs and shrubs—NS Canada	Sweeps	Percent of all ladybirds	1990–1996 >12.5% [a]	1998 0.3%	*Coccinella septempunctata, Hippodamia variegate, Propylea quatuordecimpunctata*
Schellhorn (1998)[b] (earlier), J.P. Harmon *unpubl. data*[b]	Sweet corn—St. Paul, MN USA	Visual search	Adults/plant	1994 0.28 1996 0.44	1999–2000 0 on 960 plants	*Coccinella septempunctata, Harmonia axyridis* (later)
Wold et al. (2001)[b] (earlier), R. L. Koch *pers. comm.* (later)	Sweet corn—Rosemount, MN USA	Visual search	Adults/plant	1998 0.85 1999 0.26	2000–2004 1 on 3978 plants	*Coccinella septempunctata, Harmonia axyridis*

"Before" and "after" columns contrast populations before and after adventive coccinellid(s) arrival. "Adventive coccinellids" are present in both sets of samples unless noted

[a]Within reference is a claim that *A. bipunctata* and another native were 25% of all coccinellids with *A. bipunctata* more common then the other

[b] Samples were averaged across experimental treatments that were each representative of conditions observed in unmanipulated plots

in conspecific egg predation (Hemptinne et al. 2000), that does not seem to be the case against *H. axyridis* (Sato and Dixon 2004) which seems to be well protected against native coccinellids (Cottrell 2004; Sato and Dixon 2004). Moreover, egg cannibalism may be enhanced in systems like maize where *A. bipunctata* lays most of its eggs near large aphid aggregations (Schellhorn and Andow 1999), which are also the aggregation sites of *H. axyridis* adults and later instar larvae (*J.P.H. unpublished data*). The recent findings of *A. bipunctata* in Japan (Sakuratani 1994; Toda and Sakuratani 2006) and *H. axyridis* in the United Kingdom (http://www.harlequin-survey.org/) will allow researchers to look at these and other possible interactions in new environmental settings.

It is not clear how *C. septempunctata* and *H. axyridis* have interacted in regards to their possible effects on *A. bipunctata*. As mentioned, *C. septempunctata* may have been completely responsible for observed population declines. Alternatively, *A. bipunctata* may have been negatively affected by *C. septempunctata* before *H. axyridis* arrived, only to have these effects counterbalanced or replaced by negative effects from *H. axyridis* once it arrived. This could happen, for example, if *H. axyridis* is suppressing *C. septempunctata* (as suggested in Brown 2003). Further research on combined or offsetting effects between these two exotic species may help untangle these possibilities.

Coccinella novemnotata

Evidence for decline

Coccinella novemnotata, hereafter C-9 was believed to be the most commonly collected member of the family Coccinellidae in the Northeastern US (Gordon 1985). According to the literature and field guides, C-9 was found throughout the continental United States and southern Canada (Gordon 1985) and was recognized for its considerable economic importance as a predator of many of the more destructive aphid species (Pack 1925). By the mid-1990s it became obvious that C-9 was no longer as common as everyone assumed that it had been in the 1980s and probably the 1970s. The historically broad geographic range and prominence of C-9 stands in stark contrast to its current distribution (Table 3). Besides published reports, a search of major insect collections in the Northeastern US as well as USDA records revealed that C-9 was not common in the 1980s or early 1990s (Stephens 2002). Based on the latest records in the literature, C-9 was last collected in Maryland in 1986, Pennsylvania in 1987, Delaware in 1988 and Maine in 1992 (Wheeler and Hoebeke 1995).

Table 3 Surveys of *C. novemnotata* through time

Time period	Average relative abundance of C9	Total # of studies	# Of studies with C9	Average rank of C9[a]
1914–1964	6.6	13[b]	11	3.1
1973–1989	0.06	8	6	6.5
1990–2006	0	32	0	na

Data here include studies from the full data set that were conducted in areas within C9 published range as well as data from an additional study (Ellis et al. 1999)

[a]Average rank in abundance of C9 among all coccinellids in a system, only includes studies where C9 was present

[b]5 of 13 references are from a single reference across multiple habitats

As far as recent collections go, C-9 has been collected in the Midwest and West Coast but only sporadically in these areas. Near Madison, Wisconsin, two C-9 were collected the summer of 2000 on milkweed eating milkweed aphids, *Aphis nerii*. In Washington State near Pullman, several maculate C-9 were collected the summer of 2001. The spotless or immaculate variety of *C. novemnotata* that is the primary phenotype of C-9 on the West Coast was collected in the Pajaro Valley near Santa Cruz, California.

From dissertations at Cornell University, one can reconstruct the coccinellid fauna composition around Ithaca, NY in particular, with an additional data point for Long Island, NY. In a sample taken in 1924 in alfalfa in Ithaca, NY C-9 made up 13% of the Coccinellidae collected (Pack 1925). A survey of natural enemies in potato on Long Island, NY from 1956 to 1958 found that C-9 made up 19% of the Coccinellidae collected (Day 1965). By 1971 C-9 was evidently less common in Ithaca, NY in alfalfa and represented by a maximum of only 1 individual counted per weekly 100 stem counts (Wheeler 1971).

Potential reasons for decline

Although the critical time for taking data on this disappearance is obviously past, it is important that we garner what we can from the plight of C-9 because other native coccinellids may also be vulnerable to extinction. As with many rare and endangered species, often extinction has a primary cause like habitat destruction or degradation, but many factors actually act in concert to push the species over the brink. Data can be used to gain insight into two important hypotheses for C-9's disappearance; the succession of farmland and the widespread establishment of adventive species.

The Northeast US has undergone an unprecedented greening in the last 120 years precipitated by succession of agricultural land on a large scale. Although the succession of land is often good for many species, it may mean less favorable foraging for C-9 which may have favored agricultural habitats (Gordon 1985, but see Gagne and Martin 1968). Beginning in the 1880s agricultural land in New York began to be abandoned because it was difficult to farm and required more effort for sub-optimal yields in comparison to the land available in the Midwest and Western U.S. (Williams 1989). There was a 57% reduction in land in farms, a 50% reduction in cropland and a 33% reduction in land in orchards between 1940 and 1997 (U.S.D.A. N.A.S.S.; New York State Agricultural Statistics Service 2001). Also, between 1959 and 1997, the land in two major field crops was reduced drastically; wheat acreage diminished by 50% and potatoes by 68% (U.S.D.A. N.A.S.S.). While these large scale changes in habitat and prey availability suggest a possible explanation, the critical data necessary to demonstrate causality was never collected. Moreover, more recent studies have not found any correlation between a changing landscape in South Dakota and Coccinellid densities (Elliot and Kieckheffer 1990; Elliott et al. 1999)

Many entomologists assume that adventive lady beetles, such as *C. septempunctata* and *H. axyridis* played a role in C-9's disappearance, but there is a lack of studies that document a causal relationship (Wheeler and Hoebeke 1995). The timing of C-9's disappearance in Michigan suggests that *H. axyridis* was not responsible for C-9's decline because C-9 was not showing up in collections before *H. axyridis*' arrival (Colunga-Garcia and Gage 1998). Being a member of the same genus, *C. septempunctata* may have occupied a similar niche as *C. novemnotata* (Gordon 1985) resulting in overlap in the use of habitats and possibly intraguild predation events. These *Coccinella* species had similar developmental times and egg and pupal durations and could therefore overlap temporally in their use of habitats (ElHag and Zaitoon 1996). There is a lack of data around the time of C-9's disappearance and *C. septempunctata's* establishment and no studies that determine the frequency, probability or direction of intraguild predation events. Unless more data are uncovered, it will be impossible to draw any conclusions about adventive species causing C-9's disappearance or of the possibility of other alternative hypotheses.

Summary and conclusions

From our survey of the published literature, it is clear that in just the last 20 years coccinellid assemblages of Canada and the United Stated have shifted from being all native to having some, often times very large, proportion of adventive coccinellid species. Less clear is what effect adventives have had on coccinellids as a whole, their ecological function, and why some systems and species have been more affected by adventives than others.

The few studies that have encompassed a coccinellid invasion demonstrate that the number of natives and the total number of coccinellids have sometimes increased and sometimes decreased following the establishment of adventives (Table 1). While the additional, larger survey is suggestive of a strong directional trend, the use of proportions can sometimes be difficult to interpret. Specifically, some or all of the observed decrease in the proportion of native species could simply be due to the addition of adventive species to the assemblage and not a change in the actual density of natives (as in Elliott et al. 1996 alfalfa & grains). However, our review found that the average proportion of native coccinellids decreased by 68% after 1986. To find this large a decrease in the proportion of natives by only adding adventive coccinellids, adventives would have had to increase the total abundance of coccinellids by more than 50% while having no effect on the number of natives. There is one case where such a large number of adventives were added to the assemblage (Alyokhin and Sewell 2004), yet in this system the large insurgence of adventives coincided with a 72% decrease in the density of natives. It is therefore unlikely that the entire decline in the proportion of native species seen in many assemblages has been due to the simple addition of adventive species. Given the tremendous variation in the proportion of natives observed across assemblages since 1987, we can speculate that there are a number of systems in which adventives had a minor or even positive effect on natives. However, it is likely that some systems, especially those with a small proportion of natives, are additional examples of a coccinellid invasion coinciding with a decrease in the native coccinellid population. Unfortunately, we can infer even less from the larger data set on how total coccinellid populations have been affected and are left to rely on the variable results reported across the small number of long-term studies.

Coccinellids, especially the aphidophagous species that dominate published surveys, are not well known for tightly coupled, evolutionarily derived relationships with specific species or habitats. Therefore, it can be difficult to predict what specific ecological functions have suffered from the diminishing of particular species. There is an obvious inherent risk to the presence of native species, but it is not clear what risks adventives are posing to the ecological services coccinellids provide. In fact there may be evidence that adventives have actually helped coccinellids regulate aphids in managed systems. Two recent studies have documented long-term declines in aphid densities since the establishment of an adventive species (Alyokhin and Sewell 2004; Evans 2004). Moreover, one adventive coccinellid, *H. axyridis*, is regarded as one of the only established natural enemies capable of regulating a major recent adventive pest in North American agriculture (Rutledge et al. 2004). More work is needed throughout native ecosystems to understand what additional risks adventives may pose to coccinellids and their functions in these less well-studied environments.

Our case studies on the decline of *A. bipunctata* and *C. novemnotata* illustrate that at least some species in some systems have recently undergone extensive declines. Additional circumstances such as changing land use practices may have helped exacerbate situations, but in both cases it seems extremely likely that adventive species played a role. What is needed now is a better understanding of why these two species have undergone more substantial changes than other natives. Are there specific characteristics in their physiology, ecology, or habitat use that put coccinellid species more at risk, or are there specific types of habitats, ecosystems, or landscapes that put any species more at risk than others? While these are not new questions to conservation or risk assessment, coccinellid ecology would benefit greatly from the application of these other disciplines' current tools and theories. Current invasions of coccinellids in Europe, Asia, and Australia highlight the need to understand what risks are posed to native coccinellids and what can be done to emasculate the negative effects of adventive species.

Acknowledgements The authors wish to thank Timothy New for the opportunity to publish within this special section. We also wish to acknowledge Rachel Clancy for her assistance in gathering data and Tony Ives for his statistical advice. JPH was funded by NSF grant EF-0313737.

References

Alyokhin A, Sewell G (2004) Changes in a lady beetle community following the establishment of three alien species. Biol Invasions 6:463–471

Angalet GW, Tropp JM, Eggert AN (1979) *Coccinella septempunctata* in the United States: recolonization and notes on its ecology. Environ Entomol 8:896–901

Bazzocchi GG, Lanzoni A, Accinelli G, Burgio G (2004) Overwintering, phenology and fecundity of *Harmonia axyridis* in comparison with native coccinellid species in Italy. BioControl 49:245–260

Boiteau G, Bousquet Y, Osborn WPL (1999) Vertical and temporal distribution of Coccinellidae (Coleoptera) in flight over an agricultural landscape. Can Entomol 131:269–277

Bosque-Perez NA, Johnson JB, Schotzko DJ, Unger L (2002) Species diversity, abundance, and phenology of aphid natural enemies on spring wheats resistant and susceptible to Russian wheat aphid. BioControl 47:667–684

Brown MW (2003) Intraguild responses of aphid predators on apple to the invasion of an exotic species, *Harmonia axyridis*. BioControl 48:141–153

Brown MW, Miller SS (1998) Coccinellidae (Coleoptera) in apple orchards of eastern West Virginia and the impact of invasion by *Harmonia axyridis*. Entomol News 109:143–151

Colunga-Garcia M, Gage SH (1998) Arrival, establishment, and habitat use of the multicolored Asian lady beetle (Coleoptera: Coccinellidae) in a Michigan landscape. Environ Entomol 27:1574–1580

Colunga-Garcia M, Gage SH, Landis DA (1997) Response of an assemblage of Coccinellidae (Coleoptera) to a diverse agricultural landscape. Environ Entomol 26:797–804

Cormier CM, Forbes TA, Jones TA, Morrison RD, McCorquodale DB (2000) Alien invasion: the status of non-native lady beetles (Coleoptera: Coccinellidae) in industrial Cape Breton, Nova Scotia. Northeast Nat 7:241–247

Cottrell TE (2004) Suitability of exotic and native lady beetle eggs (Coleoptera : Coccinellidae) for development of lady beetle larvae. Biol Control 31:362–371

Cranshaw W, Sclar DC, Cooper D (1996) A review of 1994 pricing and marketing by suppliers of organisms for biological control of arthropods in the United States. Biol Control 6:291–296

Day WH (1965) The Identification and Importance of Biotic and Abiotic Factors Affecting Aphids on Long Island Potatoes. PhD Dissertation. Cornell University, Ithaca NY USA

DeBach P (eds) (1964) Biological control of insect pests and weeds. Chapman & Hall, London, UK

Dobzhansky T (1935) A list of Coccinellidae of British Columbia. J NY Entomol Soc 43:331–336

Dowell RV, Cherry RH (1981) Survey traps for parasitoids, and coccinellid predators of the citrus blackfly, *Aleurocanthus woglumi*. Entomol Exp Appl 29:356–361

ElHag ETA, Zaitoon AA (1996) Biological parameters for four coccinellid species in Central Saudi Arabia. Biol Control 7:316–319

Elliot NC, Kieckheffer RW (1990) A 13-year study of the aphidophagous insects of alfalfa. Prairie Naturalist 22:87–96

Elliott N, Kieckhefer R, Kauffman W (1996) Effects of an invading coccinellid on native coccinellids in an agricultural landscape. Oecologia 105:537–544

Elliott NC, Kieckhefer RW, Lee JH, French BW (1999) Influence of within-field and landscape factors on aphid predator populations in wheat. Landscape Ecol 14:239–252

Ellis DR, Prokrym DR, Adams RG (1999) Exotic lady beetle survey in northeastern United States: *Hippodamia variegata* and *Propylea quatuordecimpunctata* (Coleoptera: Coccinellidae). Entomol News 110:73–84

Evans EW (2004) Habitat displacement of North American ladybirds by an introduced species. Ecology 85:637–647

Ewing HE (1914) Some coccinellid statistics. J Econ Entomol 7:440–443

Fenton FA, Howell DE (1955) A comparison of five methods of sampling alfalfa fields for arthropod populations. Ann Entomol Soc Am 50:606–611

Fluke CL (1925) Natural enemies of the pea aphid (Illinoia pisi Kalt.); Their abundance and distribution in Wisconsin. J Econ Entomol 18:612–615

Gagne WC, Martin JL (1968) The insect ecology of red pine plantations in Central Ontario. V. The Coccinellidae (Coleoptera). Can Entomol 100:835–846

Godarzy K, Davis DW (1956) Natural enemies of the spotted alfalfa aphid in Utah. J Econ Entomol 51:612–616

Gordon RD (1985) The Coccinellidae (Coleoptera) of America north of Mexico. J NY Entomol Soc 93:1–912

Gordon RD, Vandenberg N (1991) Field guide to recently introduced species of Coccinellidae (Coleoptera) in North America, with a revised key to North America genera of Coccinellini. Proc Entomol Soc Washington 93:845–864

Hemptinne JL, Dixon AFG, Gauthier C (2000) Nutritive cost of intraguild predation on eggs of *Coccinella septempunctata* and *Adalia bipunctata* (Coleoptera : Coccinellidae). Eur J Entomol 97:559–562

Hesler LS, Kieckhefer RW, Catangui MA (2004) Surveys and field observations of *Harmonia axyridis* and other Coccinellidae (Coleoptera) in eastern and central South Dakota. Trans Am Entomol Soc 130:113–133

Hesler LS, Kieckhefer RW, Evenson PD (2000) Abundance of cereal aphids (Homoptera : aphididae) and their predators in spring wheat-alfalfa intercrops under different crop management intensities. Great Lakes Entomol 33:17–31

Hodek I, Honěk A (1996) Ecology of *Coccinellidae*. Kluwer Academic Publishers, Dordrecht, The Netherlands

Hoffmann MP, Orfanedes MS, Pedersen LH, Kirksyland JJ, Hobeke ER, Ayyappath R (1997) Survey of lady beetles (Coleoptera: Coccinellidae) in sweet corn using yellow sticky cards. J Entomol Sci 32:358–369

Iperti G (1999) Biodiversity of predaceous coccinellidae in relation to bioindication and economic importance. Agric Ecosys Environ 74:323–342

Knowles TC (2006) Alfalfa aphid complex (blue alfalfa aphid, pea aphid, and the spotted alfalfa aphid). http://ag.arizona.edu/pubs/insects/az1044/

Koch RL (2003) The multicolored Asian lady beetle, *Harmonia axyridis*: a review of its biology, uses in biological control, and non-target impacts. J Insect Sci 3:32

LaMana ML, Miller JC (1996) Field observations on *Harmonia axyridis* Pallas (Coleoptera: Coccinellidae) in Oregon. Biol Control 6:232–237

Lee RE (1980) Aggregation of Lady Beetles on the Shores of Lakes (Coleoptera, Coccinellidae). Am Midland Nat 104:295–304

Magurran AE (1988) Ecological diversity and its measurement. Princeton University Press, Princeton, NJ USA

Mareida KM, Gage SH, Landis DA, Wirth TM (1992) Ecological observation on predatory Coccinellidae (Coleoptera) in Southwestern Michigan. Great Lakes Entomol 25:265–270

Musser FR, Nyrop JP, Shelton AM (2004) Survey of predators and sampling method comparison in sweet corn. J Econ Entomol 97:136–144

New York State Agricultural Statistics Service (2001) New York Agricultural Statistics 2000–2001 http://www.nass.usda.gov/ny/bulletin/2001/01-bulle.htm. Albany, NY USA.

Obrycki JJ, Elliot NC, Giles KL (2000) Coccinellid introductions: potential for evaluation of nontarget effects. In: Follet PA, Duan JJ (eds) Non-target effects of biological control. Kluwer Academic Publishers, Boston MA USA, pp 127–145

Pack HJ (1925) A biological study of certain ladybird beetles (Coccinellidae). Cornell University, Ithaca, NY USA

Pedigo L, Rice M (2006) Entomology and pest management, 5th edn. Pearson, Upper Saddle River, NJ USA

Putnam WL (1964) Occurrence and food of some Coccinellids (Coleoptera) in Ontario peach orchards. Can Entomol 96:1149–1155

Rosenheim JA (2001) Source-sink dynamics for a generalist insect predator in habitats with strong higher-order predation. Ecol Monogr 71:93–116

Rutledge CE, O'Neil RJ, Fox TB, Landis DA (2004) Soybean aphid predators and their use in integrated pest management. Ann Entomol Soc Am 97:240–248

Sakuratani Y (1994) New record of *Adalia bipunctata* (Linnaeus) (Coleoptera: Coccinellidae) from Japan. Japanese Jpn J Appl Entomol Zool 62:627–628

Sato S, Dixon AFG (2004) Effect of intraguild predation on the survival and development of three species of aphidophagous ladybirds: consequences for invasive species. Ag Forest Entomol 6:21–24

Schellhorn NA (1998) Cannibalism and interspecific predation: the interaction among Coccinellid beetles, their aphid prey, and maize. Ph.D. dissertation. University of Minnesota, St Paul MN

Schellhorn NA, Andow DA (1999) Cannibalism and interspecific predation: role of oviposition behavior. Ecol Appl 9:418–428

Smith BC (1958) Notes on relative abundance and variation in elytral patterns of some common coccinellids in the Belleville district. Ann Rep Entomol Soc Ontario 88:59–60

Smith BC (1971) Effects of various factors on the local distribution and density of coccinellid adults on corn (Coleoptera: Coccinellidae). Can Entomol 103:1115–1120

Stephens EJ (2002) Apparent extirpation of *Coccinella novemnotata* in New York State: optimizing sampling methods and evaluating explanations for decline MS Thesis. Cornell University, Ithaca NY USA

Toda Y, Sakuratani Y (2006) Expansion of the geographical distribution of an exotic ladybird beetle, *Adalia bipunctata* (Coleoptera: Coccinellidae), and its interspecific relationships with native ladybird beetles in Japan. Ecol Res 21:292–300

Turnock WJ, Turnock RW (1979) Aggregations of lady beetles (Coleoptera: Coccinellidae) on the shores of Lake Manitoba. Manitoba Entomol 13:21–22

Turnock WJ, Wise IL, Matheson FO (2003) Abundance of some native coccinellines (Coleoptera : Coccinellidae) before and after the appearance of *Coccinella septempunctata*. Can Entomol 135:391–404

U.S.D.A. N.A.S.S. 1969, 1974, 1982, 1992, 1997. Census of Agriculture- State data. Government Printing Office, Washington DC USA

Watve CM, Clower DF (1976) Natural enemies of the bandedwing whitefly in Louisiana. Environ Entomol 5:1075–1078

Wheeler AG (1971) A Study of the Arthropod Fauna of Alfalfa. PhD Dissertation. Cornell University, Ithaca NY USA

Wheeler AG Jr, Hoebeke ER (1995) *Coccinella novemnotata* in northeastern North America: historical occurrence and current status (Coleoptera: Coccinellidae). Proc Entomol Soc Washington 97:701–716

Williams M (1989) Americans and their forests: A Historical geography. Cambridge University Press

Wise IL, Turnock WJ, Roughley RE (2001) New records of coccinellid species for the province of Manitoba. Proc Entomol Soc of Manitoba 57:5–10

Wold SJ, Burkness EC, Hutchison WD, Venette RC (2001) In-field monitoring of beneficial insect populations in transgenic corn expressing a *Bacillus thuringiensis* toxin. J Entomol Sci 36:177–187

Wright RJ, DeVries TA (2000) Species composition and relative abundance of Coccinellidae (Coleoptera) in South Central Nebraska field crops. J Kansas Entomol Soc 73:103–111